绿色丝绸之路发展报告

（2023）

推进“一带一路”建设工作领导小组办公室

生态环境部

中国环境出版集团·北京

图书在版编目（CIP）数据

绿色丝绸之路发展报告. 2023 : 汉文、英文 / 推进“一带一路”建设工作领导小组办公室，生态环境部编著. -- 北京 : 中国环境出版集团，2023.9
ISBN 978-7-5111-5633-4

Ⅰ. ①绿… Ⅱ. ①推… ②生… Ⅲ. ①丝绸之路－生态环境保护－研究报告－2023 Ⅳ. ① X321.2

中国国家版本馆 CIP 数据核字（2023）第 184273 号

出 版 人 武德凯
责任编辑 黄 颖 梅 霞
装帧设计 宋 瑞

出版发行 中国环境出版集团
（100062 北京市东城区广渠门内大街 16 号）
网　　址：http://www.cesp.com.cn
电子邮箱：bjgl@cesp.com.cn
联系电话：010-67112765（编辑管理部）
发行热线：010-67125803，010-67113405（传真）
印　　刷 玖龙（天津）印刷有限公司
经　　销 各地新华书店
版　　次 2023 年 9 月第 1 版
印　　次 2023 年 9 月第 1 次印刷
开　　本 787×1092 1/16
印　　张 7.25
字　　数 100 千字
定　　价 80.00 元（全 2 册）

目　　录

前　　言

当今世界正经历百年未有之大变局，世界多极化、经济全球化、社会信息化、文化多样化深入发展，各国人民的命运从未像今天这样紧密相连。面对时代命题，2013 年秋，中国国家主席习近平在出访哈萨克斯坦和印度尼西亚时先后提出共建“丝绸之路经济带”与“21 世纪海上丝绸之路”的重大倡议。2015 年 3 月，中国发布《推动共建丝绸之路经济带和 21 世纪海上丝绸之路的愿景与行动》，明确提出共建绿色丝绸之路。2017 年 5 月，习近平主席在首届“一带一路”国际合作高峰论坛上提出，要践行绿色发展的新理念，倡导绿色、低碳、循环、可持续的生产生活方式，加强生态环保合作，建设生态文明，共同实现 2030 年可持续发展目标。2019 年 4 月，习近平主席在第二届“一带一路”国际合作高峰论坛上进一步强调，把绿色作为底色，推动绿色基础设施建设、绿色投资、绿色金融，保护好我们赖以生存的共同家园。2023 年 10 月，习近平主席在第三届“一带一路”国际合作高峰论坛上宣布了中国支持高质量共建“一带一路”的八项行动，并将“促进绿色发展”作为行动之一，为“一带一路”绿色发展明确了新方向，开辟了新愿景，注入了新动力。

推进绿色丝绸之路建设，是践行绿色发展理念、推进生态文明建设的内在要求，是积极应对气候变化、维护全球生态安全的重大举措，是推进共建“一带一路”高质量发展、构建人与自然生命共同体的重要载体。近年来，中国不断完善绿色丝绸之路顶层设计，发起了系列绿色行动倡议，实施了一批绿色发展项目，绿色丝绸之路建设迈上新台阶。

——登高望远，以顶层设计定向把舵。在推进“一带一路”建设工作领导小组统一部署下，各成员单位凝心聚力，协同建立并完善推进绿色丝绸之路建设的政策体系。自2017年起，中国政府相继发布《关于推进绿色“一带一路”建设的指导意见》《“一带一路”生态环境保护合作规划》《关于推进共建“一带一路”绿色发展的意见》等文件，明确总体要求，部署重点任务和工作安排。《中华人民共和国国民经济和社会发展第十四个五年规划和2035年远景目标纲要》《中共中央　国务院关于完整准确全面贯彻新发展理念做好碳达峰碳中和工作的意见》以及《国务院关于印发2030年前碳达峰行动方案的通知》等系列文件，也将推进绿色丝绸之路建设相关工作作为重要内容。

——心之所向，以中国力量彰显担当。截至2023年11月，中国已与150多个国家和30多个国际组织签署的200多份共建“一带一路”合作文件中，共建绿色丝绸之路是其中的重要内容。“一带一路”绿色发展国际联盟已汇集中外合作伙伴150余家，举办绿色丝绸之路建设主题活动70余场，传播范围超过160多个国家和地区。《“一带一路”绿色投资原则》涉及的机构已包括来自17个国家和地区的46家签署机构和17家支持机构，持有或管理的资产超过41万亿美元。中国政府组织开展的350余期环境保护和应对气候变化培训班及学历学位项目为有关国家培训专业人才1.1万余名。其中，绿色丝路使者计划和“一带一路”应对气候变化南南合作计划为120多个发展中国家培训3000余名环境与气候领域的官员和技术人员。

——脚踏实地，以务实合作共谋发展。当今世界，绿色低碳转型潮流浩浩荡荡。近年来，美丽中国建设迈出重大步伐，绿色发展成为中国式现代化的显著特征，既造福了中国，也造福了世界。根据中国国家能源局发布的数据，2022年，中国出口的风电和光伏产品为其他国家减排二氧化碳约5.73亿吨，合计减排 28.3 亿吨，约占全球同期可再生能源折算碳减排量的 41%。同时，中国在太阳能发电、风电、水电、热能等领域与全球 100 多个国家开展了合作，为共建国家能源供给向高效、清洁、多元化方向加速转型注入了强劲动力。

“一带一路”倡议源于中国，但机会和成果属于世界。绿色低碳、可持续发展关乎全人类的共同命运，关乎当代和子孙后代的利益福祉，需要国际社会汇聚“和合”之力。展望未来，绿色丝绸之路建设顺应全球绿色低碳发展的大潮流，顺应各国人民对优美环境的新期待，为共建“一带一路”国家落实联合国 2030 年可持续发展议程提供经验、路径和方案，在新形势下将展现出更加强大的生机和活力。同心前行，路致远方，中国将携手共建“一带一路”国家，厚植高质量发展的绿色底色，坚持人与自然和谐共生理念，践行共商共建共享原则，以高标准、惠民生、可持续为目标，持续提升绿色发展合作水平，建设更紧密的绿色发展伙伴关系，共同构建人与自然生命共同体，谱写高质量共建“一带一路”新华章。

一、发展成效

在中国与共建“一带一路”国家的携手努力下，绿色丝绸之路建设成果丰硕，推动了绿色基础设施“硬联通”、绿色规则标准“软联通”、绿色交流合作“心联通”，探索出了经济发展与环境保护“双赢”的新路径，绿色成为高质量共建“一带一路”的鲜明底色，为共建“一带一路”国家实现绿色低碳转型和可持续发展注入了强大动力。

（一）凝聚绿色发展国际合作共识

1. 深化绿色发展政策沟通

——加快绿色发展战略对接。截至 2023 年 11 月，中国已与 150 多个国家和 30 多个国际组织签署了 200 多份共建“一带一路”合作文件，共建绿色丝绸之路是其中的重要内容。中国会同 28 个国家发起“一带一路”绿色发展伙伴关系倡议，与来自 21 个国家的政府与环境主管部门、国际组织等发起《“一带一路”绿色发展北京倡议》，与 34 个国家共同发布《数字经济和绿色发展国际经贸合作框架倡议》，与联合国环境规划署签署《关于建设绿色“一带一路”的谅解备忘录（2017—2022 年）》，有关共识被纳入重要双边文件以及中国同东盟、中亚五国、中东欧国家、太平洋岛国合作及中阿合作论坛、中非合作论坛、中国—拉共体论坛等合作机制的成果文件。中国与伊朗、白俄罗斯、巴基斯坦等国环境与气候主管部门签署环境保护、可持续发展、应对气候变化等领域的合作文件，与俄罗斯、匈牙利、吉尔吉斯斯坦、阿根廷、塞尔维亚、新加坡、印度尼西亚、巴布亚新几内亚等国政府签署绿色发展领域投资合作谅解备忘录，不断拓展清洁能源、节能减排、绿色技术创新等领域合作空间。

——构建重点区域合作机制。中国与共建国家在联合国、二十国集团等框架下开展绿色交流合作，通过联合国环境大会、二十国集团环境部长会议、中国—东盟应对气候变化与生态环境对话等多双边合作机制开展对话交流。成立中国—上海合作组织环境保护合作中心、澜沧江—湄公河环境合作中心和中非环境合作中心，成为相关政府部门、国际组织、研究机构、企业等多方参与的重要平台。

2. 搭建绿色交流合作平台

——搭建政策对话和信息共享平台。在第二届“一带一路”国际合作高峰论坛期间，中外合作伙伴共同发起“一带一路”绿色发展国际联盟（简称绿色联盟），启动“一带一路”生态环保大数据服务平台。截至2023年8月，绿色联盟已有来自43个国家的150余家合作伙伴，包括26个共建国家的环境主管部门，成为绿色丝绸之路框架下首个国际性社会团体，迄今已举办“一带一路”绿色创新大会等70余场专题研讨会、发布《“一带一路”项目绿色发展指南》等20余份政策研究报告，发起绿色发展投融资合作伙伴关系，与“一带一路”媒体传播联盟共同启动“绿色丝路行”国际传播活动。“一带一路”生态环保大数据服务平台已汇集60多个共建国家的基础环境信息、环境法律法规和标准，囊括30多个国际权威平台公开的200余项指标数据，完成大气环境、生物多样性、水资源和水环境等专题指标数据集成。

——搭建务实合作领域专业平台。在能源领域，中国于2019年4月与有关共建国家共同发起成立的“一带一路”能源合作伙伴关系是能源领域首个由中方发起成立的国际合作平台，截至2023年11月，成员国已达到33个。迄今已召开两届“一带一路”能源部长会议和三届伙伴关系论坛，并发布《“一带一路”绿色能源合作青岛倡议》《“一带一路”能源合作伙伴关系章程》以及能源国际合作最佳实践等多项成果。在城市可持续发展领域，中国国家发展改革委与联合国人居署、世界卫生组织和能源基金会共同发起成立“一带一路”可持续城市联盟，联盟连续举办了四届国际可持续城市发展高层论坛，发布《绿色城镇化共识》，形成系列研究报告，为27个共建国家、190名学员分享中国城市可持续发展经验和案例。在绿色金融领域，中国金融学会绿色金融专业委员会与伦敦金融城于2018年11月联合发起《“一带一路”绿色投资原则》，现已包括来自17个国家和地区的46家签署机构和17家支持机构。

3. 推动绿色发展能力建设

中国积极支持共建国家能力提升，加强环境管理人员和专业技术人才互动交流，为绿色发展提供人才支撑。如实施绿色丝路使者计划和“一带一路”应对气候变化南南合作计划，举办近百场环境与气候变化领域培训活动，累计为120多个发展中国家提供3000余个培训名额，启动实施“一带一路”生态环保人才互通计划，设置环境管理与可持续发展等学历学位项目，为有关国家培训1.1万余名专业人才，得到参与国、国际组织以及社会各界的广泛认同。中国相关机构结合不同地区实际需求有针对性地开展能力建设活动，实施“中非绿色使者计划”，举办支持“非洲绿色长城”建设研修班；连续多年开设“雨水集蓄利用技术国际培训班”，派遣专家到约旦、沙特、尼日利亚、南非等国开展培训，得到联合国及相关行业协会等国际组织的高度认可。

（二）积极推进绿色基础设施建设

1. 打造绿色基础设施

2021年7月，为推动对外投资合作践行绿色发展理念，中国政府发布《对外投资合作绿色发展工作指引》，提出建设绿色基础设施，打造绿色境外经贸合作区，支持绿色技术创新，推动企业主体绿色转型，鼓励和引导“走出去”企业增强绿色发展意识。2022 年 1 月，中国政府发布《对外投资合作建设项目生态环境保护指南》，聚焦对外投资合作建设项目整个生命周期的生态环境管理，进一步规范企业对外投资合作中的环境保护行为。中国企业在参与项目建设过程中，积极践行绿色低碳理念，用实际行动诠释了绿色是共建“一带一路”的底色，为项目所在国经济社会发展和生态环境保护做出了积极贡献。

专栏一　助力共建国家建设绿色基础设施

（一）蒙内铁路

蒙内铁路是中国和肯尼亚共建“一带一路”的旗舰项目，由中国企业承建，于 2017 年 5 月建成通车。蒙内铁路全长 471 千米，连接肯尼亚首都内罗毕和东非第一大港蒙巴萨港，是首条在海外全部采用“中国标准”建造的铁路，在非洲留下了“一带一路”绿色印迹。铁路设计充分考虑野生动物生活习性和迁徙路径，全线共设计了 14 个大型野生动物通道、79 座桥梁，高度均达 6.5 米以上，保证了长颈鹿等大型动物顺利通过，同时设置 100 多个涵洞，为动物提供饮用水水源和迁徙通道。此外，项目专门划定生态保护红线，保护当地红树林湿地公园。蒙内铁路日均开行 9 对（18 列）货物列车，使用的柴油量远低于公路运输，使温室气体排放量降低了 75%。

（二）加纳特马新集装箱码头工程项目

加纳特马新集装箱码头工程项目位于非洲西部几内亚湾，是加纳目前最大的水上工程项目，项目由中国企业承建，2020 年 6 月完工。加纳是海龟的重要繁殖基地，项目在开工前进行环境基线调查，开展环境影响评价，制定了施工期环境管理方案，确保健康、安全、环境和社会责任等各方面满足世界银行国际金融公司和加纳环保局的要求。项目专门建造了“海龟孕育中心”，加强对项目所在区域海龟就地保护，通过与加纳野生动物保护协会合作，聘请海龟孵化专家进行指导，截至2021年6月，累计孵化并放生海龟超1.7万只，孵化率78.8%（自然条件下海龟蛋孵化率约为10%～20%）。这些措施对于保护物种、维持生态平衡发挥了重要作用。

2. 建设绿色低碳工业园区

在合作建设经贸合作园区过程中，中方坚持高标准规划发展目标，高规格搭建管理体系，积极践行绿色发展理念，加强环境基础设施建设，推进产业园区污水、垃圾等集中处理与循环再利用，努力减少碳排放。加强科技创新和产业创新对接，通过制定相关质量标准和行业技术规范，推动行业企业采用国际先进标准。同时，通过引进光伏发电、新能源汽车、清洁能源装备、垃圾焚烧发电、废弃物回收利用等绿色产业和项目，促进产业园区相关产业循环链接，推进资源循环利用，不断提升园区绿色低碳循环发展水平。

专栏二　助力共建国家建设绿色低碳工业园区

（一）泰中罗勇工业园

中国与泰国共建的泰中罗勇工业园明确要求企业安装污水处理设施并设定排放标准，同时要求项目组关注项目对当地社区和景观的影响，建立相应的申诉和响应机制。作为引进绿色产业的典型代表，园区将环境达标作为企业入园的前置条件，有毒有害、高污染项目列入排除清单，从源头上杜绝可能对环境造成影响的项目入园，同时鼓励生物科技、新能源、新材料等行业企业入驻，要求其对标ISO14001环保标准进行厂房建设，接受第三方污水处理、监督和评估。

（二）印度尼西亚青山工业园

中国与印度尼西亚共建的青山工业园规划建设太阳能和风能等清洁能源基地，为电池等零部件生产提供清洁能源。园区利用当地丰富的镍矿储量，以“镍铁+不锈钢”一体化生产为主导产业，统筹产业发展和环境保护。通过采用先进技术，减少生产每吨镍的二氧化碳排放量60%以上。青山工业园还采取海洋垃圾治理、珊瑚移植、红树林保护等措施，使当地热带鱼和珊瑚礁数量持续增加，最大限度地减少项目对当地生态环境和生物多样性的影响，该项目被印度尼西亚政府授予“工业园区新秀奖”，成为两国产业合作示范项目。

3. 建设环境基础设施

污水和固体废物处理是发展中国家在快速工业化、城镇化进程中面临的突出环境挑战。中国企业立足当地实际需求，按照绿色低碳、集约高效、循环发展的原则，通过支持共建污水集中处理设施、固体废物处置设施，采用中国先进技术等方式帮助共建国家完善环境基础设施建设，不断提升环境基础设施绿色化、智能化水平，实现资源合理利用、污染物有效处置，有效改善当地人居环境，实现经济、社会和环境协同发展。

专栏三　助力共建国家建设环境基础设施

（一）污水治理设施建设

中国与蒙古国共建乌兰巴托中央污水处理厂，项目建成后将成为蒙古国单体规模最大、最现代化、工艺设计最复杂的污水处理厂。该项目将惠及蒙古国44%的人口，有效解决乌兰巴托市污水处理能力严重不足的问题，保障民众用水安全。中国与孟加拉国共建的达舍尔甘地污水处理厂是该国第一座现代化大型污水处理厂，同时也是南亚地区最大单体污水处理厂，每天可处理首都达卡市区近500万人口的生活污水，大幅改善了达卡水环境质量。中国与巴布亚新几内亚共建科科坡污水处理项目，涵盖污水管道、污水处理及道路绿化等工程，通过氧化、过滤、杀菌、消毒、检测等工序使污水达标排放，显著改善了当地环境。此外，2023年，中方积极推动援乌兹别克斯坦锡尔河州污水处理厂项目探讨和立项。

（二）固废处理设施建设

由中国企业投资、建设和运营的芹苴垃圾焚烧发电项目是越南第一个现代化、采用高标准建设的垃圾焚烧发电项目。该项目采用中国先进技术，推动固体废物减量化，极大促进了垃圾焚烧发电领域技术发展，并于2018年获得芹苴市优秀企业奖。位于斯里兰卡商业首都科伦坡的固体废物处理项目采用先进膜处理技术，经处理后的垃圾渗滤液可达到该国一级排放标准，直接用于普特拉姆地区的绿化灌溉，实现了固废循环化、资源化利用。中国和泰国企业联合开发的曼谷农垦垃圾发电项目自2016年3月并网发电以来一直稳定运行且实现达标排放，并通过国际标准化组织（ISO）管理体系认证，中资企业在项目过程中展现的环境社会责任受到当地政府高度认可。

（三）持续深化绿色能源交流合作

1. 共建可再生能源项目

近年来，中国可再生能源实现了跨越式发展，中国成为全球可再生能源领域的引领者和贡献者。根据国际可再生能源署的数据，截至 2022 年年底，全球可再生能源总装机容量达到 3.37 太瓦（TW），2022 年几乎一半的新增装机容量都在亚洲，其中，中国的贡献量最大，新增装机容量达 141 吉瓦（GW）。根据中国国家能源局发布的数据，2022 年，中国可再生能源发电量相当于减少国内二氧化碳排放约 22.6 亿吨，出口的风电和光伏产品为其他国家减排二氧化碳约 5.73 亿吨，合计减排 28.3 亿吨，约占全球同期可再生能源折算碳减排量的 41%。同时，中国在太阳能发电、风电、水电、热能等领域与全球 100 多个国家开展了合作，为共建国家能源供给向高效、清洁、多元化方向加速转型注入了强劲动力。埃塞俄比亚阿达玛风电项目、厄瓜多尔芦苇桥风电项目、阿根廷胡胡伊省高查瑞光伏电站项目、黑山莫祖拉风电项目、越南九安风电项目等一大批清洁、高效、优质的可再生能源项目相继建成，有效解决了当地用电难、用电贵问题，对于优化能源结构、减少污染物和温室气体排放、带动经济发展做出了积极贡献。

专栏四　助力共建国家建设可再生能源项目

（一）巴基斯坦卡洛特水电站项目

巴基斯坦卡洛特水电站项目是中巴经济走廊首个水电项目，于 2022 年 6 月正式投入运营。卡洛特水电站总装机 72 万千瓦，年平均发电量 32 亿千瓦·时，将满足当地约 500 万人口的用电需求，预计每年可节约标准煤140万吨。卡洛特水电站在建设过程中采用世界银行国际金融公司的环境和社会责任框架及相关标准，以国际标准严格把控项目环评工作，巴基斯坦前总理阿巴西评价该项目是利国利民的典范项目。项目严格落实绿色发展理念，动工前，在影响区深入开展生态环境保护宣传教育，制定明确的鱼类资源保护规范以维持大坝下游适宜的水生生物栖息地。建成后，卡洛特项目预计每年可减少二氧化碳排放量350万吨。

（二）黑山莫祖拉风电项目

黑山莫祖拉风电项目是黑山可再生能源标杆工程，可帮助其实现“新能源装机容量翻番”的目标。项目每年提供清洁电力1.1亿千瓦·时，约占黑山全国发电总量的5%，每年可减少约3 000吨温室气体排放，在提高可再生能源比例、促进减排环保效应方面做出了重要贡献。此外，该项目为黑山带来了显著社会效益，项目完工后可满足10万当地居民用电需求，极大缓解了旅游旺季经常断电的窘境，当地也因能源环境改善吸引外商投资，增加了就业机会。该项目采用了先进的“智能风电”技术，发电效率提升15%～20%，特别是配套使用中方开发的智慧风场软件，使运营成本降低了20%。

（三）阿根廷胡胡伊省光伏发电站项目

阿根廷胡胡伊省光伏发电站项目是南美地区装机容量最大、海拔最高的光伏电站项目，总装机容量约300兆瓦，设计年平均上网电量约6.6亿千瓦·时，每年可节约标煤约20.2万吨，可减排55.2万吨二氧化碳、106.1吨二氧化硫、118.7吨氮氧化物、21.2吨烟尘。该项目于2020年9月正式并网发电，2021年4月正式进入运营期。2021年，该项目累计发电 7.11 亿千瓦·时，有力缓解了当地电力供应紧张的状况，带动了当地工业和旅游业等相关产业的发展，并对提高就业水平、推动经济社会发展起到了积极促进作用。

（四）越南九安风电项目

越南九安风电项目位于越南嘉莱省九安乡，是当地首批并网的风电项目，为嘉莱省风电工程树立了新标杆。该项目 90% 以上采用中国设备，总装机容量 46.2 兆瓦，每年能提供清洁电力 1.1 亿千瓦·时，可减少二氧化碳排放量约 9 万吨，有助于越南在 2050 年前实现净零碳排放目标。该项目进一步优化了当地能源结构，提高了可再生能源比例，所生产的绿色电力可满足约 5 万当地居民用电需求，有效缓解了当地电力供应紧张的情况。该项目建设和运营期间为当地提供就业岗位约 80 个，对当地就业和经济发展起到了积极促进作用。

（五）阿联酋迪拜光热光伏混合发电项目

阿联酋迪拜 950 兆瓦光热光伏混合发电项目由 3 个装机容量 200 兆瓦的槽式电站、1 个装机容量 100 兆瓦的塔式电站和 1 个装机容量 250 兆瓦的光伏电站组成，是全球迄今为止规模最大的光热光伏混合电站，同时也是世界上装机容量最大、投资规模最大、熔盐储备热量最大的光热项目。该项目建成后将拥有世界上最大的储热能力，能够为迪拜 27 万多家住户提供清洁电力，每年减少 140 万吨碳排放。

2. 推动能源低碳转型升级

——**不再新建境外煤电项目。**中国参与的境外能源行业投资和建设项目中的化石能源占比自 2016 年开始不断下降，到 2020 年，可再生能源（水能、光伏、风能和生物质能）投资和建设项目占能源总投资比例首次超过化石能源。

2021年9月21日，习近平主席在第七十六届联合国大会一般性辩论上宣布，中国将大力支持发展中国家能源绿色低碳发展，不再新建境外煤电项目。这是继习近平主席宣布中国力争2030年前实现碳达峰、2060年前实现碳中和后，中国为应对全球气候变化自主采取的又一重要举措，也是中国为促进全球绿色复苏主动做出的又一积极贡献，展示了构建人与自然生命共同体的坚定决心和实际行动，得到了国际社会的高度认可和普遍赞誉。

——**推动重点领域节能合作**。部分共建国家正处于工业化加速发展阶段，其资源能源消耗和污染物排放处于较快增长阶段，钢铁、水泥、石化等高耗能产品需求旺盛。中国企业助力共建国家采用高效低耗技术、升级能源系统、引入大数据分析技术等措施，实现了污染减排和节水节电。奇姆肯特炼厂是哈萨克斯坦三大炼厂之一，始建于20世纪70年代，经中国企业实施环保升级改造后，污染物排放减少了90%，每年节水约150万立方米。中冶赛迪集团总承包建设的越南台塑河静钢铁项目采用高效低耗特大型高炉关键技术群，以及中国自主研发的高炉智能化生产管理系统，将大数据分析技术与实际生产应用结合，实现了钢铁绿色节能生产，主要技术指标达到了国际先进水平。

3. 加强能源技术装备合作

科技创新是实现发展方式绿色低碳转型的重要突破口。作为全球最大的绿色能源市场和装备制造国，中国已建立涵盖全产业链条的绿色能源产业，具备较强研发、制造、安装和运维能力，风电、光伏发电等设备生产规模居世界第一，光伏组件、风力发电机、齿轮箱等关键零部件占全球市场份额的70%。中国通过加强绿色能源对外投资、环境技术援助和合作，大力推动绿色能源装备、技术和服务“走出去”，助力共建国家和地区能源绿色转型。晶科能源2015年在马来西亚槟城投资建设太阳能电池片和组件生产厂，现光伏电池已达到年产能7吉瓦、组件已达到年产能7吉瓦。除制造工厂外，晶科能源在马来西亚还设立了中国光伏企业最大的境外产品研发中心，从事光伏组件开发、试验和测试工作，促进了中马先进光伏产品和技术交流。

（四）积极助力绿色低碳交通运输

1. 建设绿色交通基础设施

2023 年 9 月，中国交通运输部与联合国签署了推动可持续交通促进可持续发展目标的合作谅解备忘录。中国企业在共建国家实施的铁路、公路、机场等交通基础设施建设项目，将低碳环保理念贯穿到项目规划建设和运营的全过程。在勘察设计阶段，合理避让、减缓乃至消除对各类保护区和生态敏感脆弱区的影响，注重节能降噪、环境治理、资源集约共享等技术运用，推动交通基础设施标准化、智能化，推广永久设施和临时设施结合施工，推进建设养护一体化，降低全生命周期能耗和碳排放。以中老铁路为例，中国企业高标准落实生态保护措施，从勘察选线到施工建设，充分考虑工程可能产生的环境影响，将铁路与大自然巧妙地融为一体，将沿线民族特色和文化遗产保护与绿色交通发展理念结合。项目在建设过程中充分吸收当地社区、专家和环保人士意见，让中国企业的绿色行动真正获得当地民众认可。

专栏五　助力共建国家建设交通基础设施绿色项目

（一）中老铁路项目

中老铁路项目在建设过程中，高度重视生态环境保护，严格落实环保措施，打造绿色生态铁路。该项目积极开展环境影响评价工作，委托老挝本地公司结合项目实际编制环评报告。推动绿色长廊建设，按照“试点先行、样板引路”“四季常绿、乔灌草结合”原则，在全线选取 7 处绿化工程试验段，比选出适合当地的植被，全线绿化总面积约 307 万平方米，栽植灌木约 863 万株，乔木约 5.5 万株。落实水资源保护，全线隧道施工洞口共设置沉淀池 157 座。针对跨河桥梁钻孔桩产生的泥浆，设置泥浆储存池和沉淀池，以便循环利用。

（二）中巴喀喇昆仑公路（塔科特 - 雷科特）改线项目

该项目以“安全、畅通、环保、节约”为总目标，在开展技术标准、选线定线、建设方案比选及沿线设施设置研究时，充分考虑环境保护、资源节约、全寿命周期成本等要求。该项目在保证通行能力和安全的前提下，与地形、地质条件相适应，使公路最大程度融入自然环境。注重生态选线，采用桥梁跨越或隧道绕避生态环境敏感区。该项目重视环境影响评价工作，由巴基斯坦国家工程咨询公司（National Engineering Services Pakistan Limited, NESPAK）结合项目实际编制环评报告，项目严格执行巴基斯坦环保相关法律法规及环评要求，可行性研究报告建设方案充分吸收巴方政府、当地社区和环保人士意见，让中国企业的绿色行动真正获得当地民众认可。

（三）多哥洛美纳辛贝—埃亚德马国际机场现代化改造工程项目

多哥洛美纳辛贝—埃亚德马国际机场现代化改造工程项目被誉为西非地区的“样板工程”，曾于 2019 年获评中国对外承包工程商会“中国境外可持续基础设施项目”奖项。项目建设过程中，项目方高度重视机场可持续发展，严格落实相关生态环保要求，致力于打造绿色可持续机场。项目方注重从设计、施工、运营各个环节减少温室气体排放，保护工程所在地生态环境，如在航站楼公共区域照明控制设计上采用独立的 ABBi-bus 控制系统，减少室内灯光热量产生。践行绿色采购原则，严格选择通过 ISO14000 认证的分包方及材料设备供应方进行合作，建立了环保建材目录和淘汰及限制使用建材目录，绿色采购占总采购金额的比例达到 95%。

2. 助力绿色低碳出行

近年来，中国新能源汽车产业在行业标准、产业联盟、企业布局、技术研发等方面取得明显进展，从拉丁美洲到欧洲，从亚洲到非洲，越来越多的中国新能源汽车项目落地，助力共建国家打造智能化绿色公共交通，加快形成绿色低碳生产生活方式。2017 年，比亚迪在欧洲兴建的第一座电动大巴工厂在匈牙利投产，每年可生产约 400 辆零排放纯电动大巴，与匈牙利共同建设兼具经济和环保效益的公交系统。截至 2022 年年底，比亚迪向智利交付纯电动公交车 435 辆，超过该国纯电动公交车市场的 60%，预计 2023 年年底前交付 800 台，助力智利 2040 年前实现公交车队 100% 电动化。

3. 打造绿色物流运输链

中国与共建国家合作建设的中欧班列、西部陆海新通道班列等物流运输链，有效助力节能减排降碳，成为联通亚欧大陆的绿色贸易大通道。中欧班列作为国际陆路运输的新型组织方式，通过推动国际货物运输便利化，不断完善班列运输组织、提升班列运输效率、压缩全程运到期限，大大降低了运输过程中的碳排放及环境影响。班列平均碳排放量仅为公路运输的 1/7，而且还带动了多式联运快速发展，在海铁联运、公铁联运、空铁联运等多方面优化物流运力、运输流程，有效提升了物流效率，在应对全球气候变化、引领全球交通运输可持续发展中发挥了良好的示范作用。

（五）全面提升绿色金融合作水平

1. 搭建绿色金融发展平台

搭建绿色金融发展平台，鼓励和引导金融机构与企业扩大绿色投资，是绿色丝绸之路的重要支撑。《“一带一路”绿色投资原则》明确了披露环境信息、运用绿色金融工具、通过多方合作进行能力建设等七条原则，现已有来自17个国家的46家金融机构和企业参与，包括中国的国有商业银行和政策性银行，以及来自法国、德国、英国、新加坡、蒙古国、哈萨克斯坦、泰国、南非等国的大型金融机构。成员管理或持有的总资产规模达41万亿美元，已设立中亚区域办公室和非洲区域办公室。基于该原则，牵头开发了环境与气候风险评估工具箱和绿色项目库，并开展了绿色金融系列能力建设活动。

2. 提升绿色金融服务水平

——加强绿色投资政策引导。中国通过完善绿色金融政策不断提升对外投资合作项目可持续发展水平。2013年以来，发布了一系列政策文件引导和规范对外投资项目绿色发展。印发《对外投资合作绿色发展工作指引》《对外投资合作建设项目生态环境保护指南》等文件，持续更新《对外投资合作国别（地区）指南》，强化对中国企业参与共建国家绿色经济合作的指引，规范企业对外投资合作中的环境保护行为，推动对外投资合作项目绿色发展。同时，中国要求对外投资企业提高环境信息披露水平，鼓励运用绿色债券等绿色融资工具，推行环境污染强制责任保险等风险管理。

——参与共建国家绿色投融资实践。通过政策引导和金融产品创新，中国与共建国家撬动更多金融资源配置到绿色低碳领域，形成了一批典型项目，促进了共建国家产业绿色低碳转型升级。如中国工商银行为哈萨克斯坦扎纳塔斯风电项目提供1 300万美元融资支持，这是哈萨克斯坦首个由商业银行提供融资支持的可再生能源项目，也是其首次尝试通过项目融资结构为风电工程进行融资。中国银行作为全球牵头协调银行，成功筹组由欧洲复兴开发银行、亚洲基础设施投资银行、欧佩克国际开发基金等多家跨国银行和多边机构组成的国际银团，为总投资超过10亿美元的乌兹别克斯坦锡尔河1 500兆瓦燃气电站项目落地提供了重要的资金保障。

3. 深化绿色金融国际合作

——**加强与多边金融机构合作**。中国与亚洲基础设施投资银行、世界银行、亚洲开发银行、新开发银行、多边开发融资合作中心等多边平台建立了良好的合作关系，合力为共建国家基础设施互联互通和绿色发展提供资金支持。如推动亚洲基础设施投资银行对标国际的政策标准制定《环境和社会框架》《能源战略》等，指导投资者在能源、交通、城市等基础设施领域项目中推广绿色理念和技术。截至2023年6月，亚洲基础设施投资银行批准项目总额达412亿美元，其中应对气候变化领域融资占比超过50%。加强与世界银行、亚洲开发银行合作，动员更多资金支持共建国家应对气候变化和绿色低碳发展。支持多边开发融资合作中心积极发挥多边合作协调平台作用，促进发展中国家金融机构对标环境、社会和治理等国际标准规则，推动各国金融机构加强交流合作，助力共建“一带一路”绿色发展。

——**推动绿色金融标准国际合作**。2018年，中欧双方共同发起了可持续金融国际平台（IPSF），旨在深化国际合作，动员私人部门资金参与可持续投资。2022年6月，IPSF正式发布《可持续金融共同分类目录》，包含了中欧共同认可的72项对减缓气候变化有重大贡献的经济活动，加强了中欧绿色资产互认互通，为开展绿色投资提供了便利，有助于降低跨境交易的绿色认证成本。

——**加强绿色金融能力建设**。2018年以来，中国研究机构发起设立的

“绿色金融领导力项目”围绕绿色金融政策和实践，通过现场、线上学习和交流等方式，为来自70多个发展中国家约4 500名代表提供了能力建设支持，有效帮助发展中国家政策制定部门和金融机构提升绿色金融相关能力。中国金融机构积极支持共建国家完善绿色金融体系，如通过支持蒙古国金融机构与中国行业主管部门交流研讨，分享中国《绿色债券支持项目目录》等成果，助力蒙古国金融稳定委员会正式批准《蒙古国绿色金融分类目录》，有效推动其金融机构绿色投融资相关业务发展。

（六）持续推进绿色发展科技创新

1. 加强绿色技术研发共享

——**建立绿色技术合作机制。**发挥“一带一路”科技创新行动计划等机制作用，启动建设绿色技术领域联合实验室，启动实施“一带一路”可持续发展技术专项合作计划、“一带一路”可持续发展科学行动计划，与共建国家在绿色技术领域开展联合研究，大力提升绿色丝绸之路科技支撑能力。推动绿色领域技术合作示范，支持与巴基斯坦开展防灾减灾、水资源保护、可持续发展等领域技术研究与应用。为菲律宾、马来西亚、南非等碳减排提供技术研发支持。与埃塞俄比亚开展低海拔区灌丛草地项目，实施关键技术攻关。梳理涵盖大气污染防治、水污染防治、土壤污染防治、固废处理、环境监测、绿色交通等领域的绿色技术清单，为共建国家提供技术和经验借鉴。

——**探索绿色技术合作路径。**针对不同国家和地区的需求特点，形成了技术转移转化—联合实验室建设—技术人员培训为主的三位一体可持续发展模式。建成中—缅—蒙—刚绿色矿产国际联合实验室、中—泰—马可再生能源国际联合实验室、中—埃—苏绿色科技联合实验室等，依托跨国技术转移中心，促进先进绿色技术转移转化，为共建国家可持续发展注入创新动能。

建设“一带一路”绿色技术储备库，结合共建国家绿色发展现状及趋势，筛选出近 40 个有借鉴和推广价值的典型案例重点推广，推动绿色成果共享。

2. 促进绿色技术交流转移

中国持续推进环境技术交流与转移基地建设，创新合作模式，推动环境技术交流与转让，推进中国技术和服务扎根共建国家，建设一带一路环境技术交流与转移中心（深圳），打通孵化、投资、合作、转化全链条，业务范围覆盖 60 多个共建国家。打造旗舰型绿色技术供需对接平台，设立节能减排、氢能源、安全节能照明等 15 个绿色产业联合创新中心，推进中国与东盟等国家建设绿色技术交流基地，发挥环保产业集聚效应，推动绿色技术在共建国家落地。

3. 推进绿色规则标准对接

——**推动国际绿色标准制定。**中国推动国际标准化组织（ISO）、国际电工委员会（IEC）成立小水电、分布式电力能源系统、可持续电气化交通等与绿色低碳发展相关的 9 个国际标准组织技术机构，牵头制定并发布《节能量评估者选择通用指南》等 ISO、IEC 国际标准 59 项，为促进共建国家绿色标准协调奠定基础。

——**分享中国绿色行业标准。**中国现已发布碳达峰碳中和领域国家标准外文版 33 项，以及多项外文版绿色相关行业标准项目，其中，《房间空气调节器能效限定值及能效等级》国家标准得到国际广泛应用，在 2022 年年初被联合国评为全球最佳实践案例。通过发布绿色行业标准外文版，促进与共建国家绿色标准的兼容对接，推动与共建国家在产品能效标准比对、能效标准标识协调一致等方面交流与合作。

（七）积极开展应对气候变化合作

2015年，习近平主席在巴黎大会上宣布在发展中国家开展10个低碳示范区、100个减缓和适应气候变化项目、1 000个应对气候变化培训名额的“十百千”项目。2019年，习近平主席在第二届“一带一路”国际合作高峰论坛上宣布实施“一带一路”应对气候变化南南合作计划。截至2023年11月，中国已与39个共建国家签署48份气候变化合作文件，与埃塞俄比亚、巴基斯坦、萨摩亚、智利、古巴、埃及等30余个共建国家开展70余个减缓和适应气候变化项目，共同建设若干低碳示范区，帮助相关国家提升应对气候变化能力。

1. 合作建设低碳示范区

中国与老挝、柬埔寨等国共建低碳示范区，通过提供太阳能路灯、新能源汽车和便携式环境监测设备等物资，共同编制低碳示范区规划方案，以“物资 + 智力”相结合方式，促进当地绿色、低碳和可持续发展，帮助共建国家提高应对气候变化的能力。低碳示范区项目的实施，加强了中国与共建国家在应对气候变化和绿色低碳转型方面的共识，也为共建国家探索低碳发展路径提供了重要借鉴。

专栏六　助力共建国家建设低碳示范区

（一）老挝万象赛色塔低碳示范区

作为中国与共建国家建设的第一个低碳工业园区，万象赛色塔低碳示范区项目高度契合老挝政府提出的可再生和绿色能源发展规划，为老挝实现 2050 年温室气体净零排放目标提供助力，也为海外工业园区和综合开发区探索绿色低碳转型和可持续发展路径提供了可参考借鉴的模板。老挝低碳示范区首批援助物资包括 2 000 套太阳能路灯，28 辆新能源汽车，5 套环境监测设备。该示范区于 2022 年 4 月正式揭牌，预计每年可使当地减少约 1 243 吨碳排放，相当于植树超 10 万棵。中国带来的绿色可持续发展理念推动万象新城发展成为老挝乃至东盟国家低碳环保城市的典范。目前，中老双方已签署关于开展低碳示范区第二批物资援助项目的合作文件，将进一步助力示范区低碳化建设。

（二）柬埔寨西哈努克省低碳示范区

中国与柬埔寨共建西哈努克省低碳示范区，首批援助为包括 2 800 套太阳能路灯，200 套太阳能校园光伏系统，200 辆电动摩托车，10 套空气质量自动监测站，10 套便携式工业烟气监测、噪声监测、振动监测仪器在内的应对气候变化相关设备和物资，绿色光源照亮了西哈努克港，帮助柬埔寨迈出了应对气候变化的重要一步。目前，中柬环境部门已签署低碳示范区第二批物资援助项目合作谅解备忘录，中国将向柬方援助 1 500 套太阳能路灯、2 000 套户用光伏发电系统和 2 辆移动大气质量监测车等应对气候变化设备物资，为中柬西哈努克省低碳示范区建设继续提供支持。

2. 实施应对气候变化援助项目

——开展物资设备援助。中国高度重视共建国家在气候变化问题上的实际情况和重点关切，坚持量身定制、因地制宜的援助方案，在清洁能源、低碳交通、灾害预警、环境监测领域为受援国提供力所能及的物资支持。向巴基斯坦提供的户用光伏发电系统极大提升了当地居民的电力可及性，巴基斯坦总理称“中国的援助为当地民众带来真正的福祉”。援助古巴的光伏项目不仅帮助古巴解决了部分偏远地区农村居民的基本用电和照明问题，也为古巴实现温室气体减排目标提供了新动能。

——**实施工程项目援建**。除物资捐赠外，中国还援建了一批应对气候变化工程项目，包括秘鲁国家应急指挥中心和配套应急模拟体验中心、乌兹别克斯坦自动气象站示范站等，也为减缓当地气候变化影响做出重要贡献。其中，援助埃塞俄比亚的微小卫星项目是埃塞俄比亚的第一颗人造卫星，卫星采集的数据和形成的数据分析图谱可以帮助埃塞俄比亚对相关灾害风险做出快速应对，更好地缓解气候变化对农业、林业、水资源等相关领域的影响。援建突尼斯坦塔维市的两座水坝有效消减了强降雨等极端气候的不良影响。帮助贝宁维修马朗维尔防洪堤坝，提高了当地抵御洪灾的能力。2023 年 10 月，完成援乌兹别克斯坦风电站项目立项工作，目前正在组织实施。

3. 开展应对气候变化能力建设培训项目

在能力建设培训方面，中国累计举办 50 余期应对气候变化南南合作培训班，为 120 多个共建国家培训约 2 300 名气候变化领域的官员和技术人员，帮助相关国家提升应对气候变化能力。培训班主要面向共建国家应对气候变化政府官员、专家学者和技术人员，主题涵盖绿色低碳可持续发展、温室气体减排与能源转型、利用航天技术应对气候变化、“一带一路”气候融资、低碳技术及产业发展、干旱地区生态适应等主题，同时依托中国—太平洋岛国应对气候变化合作中心等平台机制，面向太平洋岛国、非洲国家举办区域性培训班，提高培训针对性。有关举措受到广大发展中国家的高度好评，多米尼加、尼日利亚等多国高度认可中国为当地培养了大批应对气候变化专业人才，体现了中国负责任大国的担当。

二、发展启示

十年踔厉奋发，十年春色满园。中国与共建国家共同做实合作基础、做强合作动力、做大合作空间、做优合作文章，推动绿色发展合作向更宽领域、更深层次、更高水平拓展，为建设清洁美丽的世界贡献了共建国家和地区的智慧和方案。通过十年努力，绿色发展理念更加深入人心，绿色合作共识不断凝聚，绿色目标要求融入共建“一带一路”各环节、全过程，降碳、减污、扩绿、增长协同推进，经济、社会、环境协调发展，绿色丝绸之路建设取得显著成效。

（一）始终坚持共商共建共享原则

坚持共商绿色发展路径。人类是一荣俱荣、一损俱损的命运共同体，面对气候变化、生物多样性丧失等全球生态环境挑战，共建国家秉持团结合作、共同发展的理念，在实现绿色低碳可持续发展中互学互鉴、互利共赢。绿色丝绸之路建设始终围绕共建国家的共同关注，坚持共商发展议题、合作路径和实践方案，持续加强绿色发展战略和政策对接，确定共建目标、方向和任务，完善政策对话、信息共享和技术合作平台，奠定了开放、包容、可持续发展的主基调。

坚持共建绿色发展项目。中国与共建国家深挖绿色发展领域合作潜能，因地制宜、科学设计，合理调配资源，着力打通各项合作落地的“最后一公里”，把握新能源、绿色产业等新兴市场机遇，不断推动绿色技术和项目在共建国家落地实施，为共建国家绿色发展提供务实可行的解决方案。创新国际合作模式，协调整合差异化资源，促进共建国家绿色发展需求与国际优势产能技术有效对接，第三方市场合作取得 1+1+1>3 的效果。

坚持共享绿色发展成果。中国紧跟时代、放眼世界，承担大国责任、展现大国担当，成为全球环境治理的引领者。中国深化与共建国家在生态环境保护、可持续发展等领域的合作，设立中国气候变化南南合作基金，宣布成立昆明生物多样性基金，推动绿色成果共享共用。中国立足共建国家绿色低碳发展需求，分享中国绿色发展理念和最佳实践，为共建国家解决突出环境问题、全面改善环境效益提供了系统性、综合性的绿色解决方案。

（二）始终坚持高标准、可持续、惠民生目标

坚持以高标准作为内在要求。高质量发展和高水平保护相辅相成、相得益彰。高水平保护是高质量发展的重要支撑，中国高度重视共建“一带一路”生态环境保护，不断提高对境外投资合作建设项目的生态环境标准要求，鼓励应用国际通行规则标准或中国更严格标准。在境外投资合作建设项目实施中，大力发展新能源和清洁能源，注重因地制宜设计环保方案和绿化工程。持续加强科技创新引领高标准，共同培育更多新的经济增长点，为共建国家在新领域新赛道合作提供新机遇。

坚持将惠民生作为根本宗旨。良好的生态环境是最公平的公共产品、最普惠的民生福祉。绿色丝绸之路建设是增强共建国家人民生态环境获得感的迫切需要，更是提升共建国家人民幸福感和安全感的重要抓手。中国重视与共建国家开展绿色民生项目，着力提升生态惠民能力，坚持以绿色发展成果反哺民生，在增加就业、改善人居环境、应对气候变化、防灾减灾等方面取得实实在在的成效，让共建“一带一路”成果惠及各参与者和相关方，实现

同共建国家人民“心联通”。

坚持将可持续作为核心要义。当前，全球经济增速放缓、气候变化等挑战更加突出，中国通过高水平环境保护，不断塑造发展的新动能、新优势，让同样面临实现发展和保护生态环境双重任务的共建国家看到了希望。同时，中国高度重视绿色丝绸之路建设的可持续性，聚焦绿色基础设施、绿色能源、绿色交通、绿色产业、绿色贸易等重点领域，发挥绿色金融保障作用，与共建国家努力走出一条更加稳健、更具活力、更为普惠的发展道路。

（三）始终坚持人与自然和谐共生理念

坚持生态优先。绿色低碳发展是当前全球发展的大方向、大趋势，也是世界各国的共识。中国与共建国家始终秉持绿色、低碳、可持续发展理念，努力推进经济社会发展全面绿色转型。绿色丝绸之路与各国绿色低碳发展目标相互呼应、相互促进，在应对气候变化、环境保护、能源转型等领域不断打造新的合作亮点。绿色丝绸之路用经验、技术和人才支持为共建国家留下了宝贵的绿色财富，中国以实际行动向国际社会宣示推进绿色丝绸之路的决心。

坚持立足国情。共建国家由于发展阶段和资源禀赋等不同，既要遵循一般规律，更要基于各国国情自主探索绿色低碳发展之路。共建国家大部分为发展中国家，推动绿色发展要循序渐进，有计划、分步骤开展合作，明确优先合作领域和事项。共建国家生物多样性普遍丰富，其中不乏全球生物多样性热点地区，中国与共建国家立足实际、科学谋划，在绿色丝绸之路框架下开展了形式多样的合作项目，为助力共建国家提升生态环保能力、保护生物多样性发挥了积极作用。

坚持系统观念。生态环境既是一个复杂、多维、客观的自然系统，也是一个涉及多方面、多领域的社会系统。当前，全球环境治理仍面临诸多挑战，共建绿色丝绸之路就是要以更宽视野、更高标准、更大力度来谋划和推进全球及区域可持续发展，更加注重对生态环境保护与经济发展内在统一性及其规律的运用，从支持共建国家解决突出环境问题入手，统筹考虑共建国家产业结构调整、污染治理、生态保护、应对气候变化，支持发展中国家在发展中降碳，在降碳中实现更高质量发展。

循道而行，功成事遂。绿色丝绸之路建设既是一项长期任务，也是一项系统工程，总结十年的实践经验，要始终保持绿色发展决心定力，坚持共商共建共享合作原则，坚持高标准、惠民生、可持续目标，统筹各领域资源，汇聚各方面力量，系统推进，久久为功，就一定能形成共建清洁美丽世界的强大合力，实现人与自然和谐共生的现代化。

三、发展展望

绿色发展是顺应自然、促进人与自然和谐共生的发展，是高质量、可持续的发展，已成为共建国家的广泛共识和共同目标。绿色丝绸之路高度契合共建国家人民对美好生活的向往，与联合国可持续发展目标要求深度对接，在新时代必将焕发出勃勃生机。行而不辍，未来可期。中国将与共建国家一道，聚同化异、精诚合作，建设更加紧密的绿色发展伙伴关系，推动政策规则标准衔接互动，促进全方位务实合作，强化绿色发展能力建设，推动绿色丝绸之路建设取得新成效，为共同构建人与自然生命共同体做出新贡献。

（一）携手提升绿色发展合作水平

1. 全面夯实绿色发展合作基础

和平、发展、合作、共赢，是不可阻挡的历史潮流。十年来，“一带一路”绿色发展理念和政策引领不断强化，绿色发展交流平台和合作机制不断完善，绿色发展伙伴关系不断紧密，绿色发展务实合作不断深化，中国与共建国家绿色发展已奠定坚实的合作基础。未来，中国愿与共建国家持续推进全方位合作，坚持践行共商共建共享原则，以高标准、惠民生、可持续为目标，以生态优先、绿色发展为导向，推动绿色丝绸之路建设与联合国可持续发展议程深入对接，促进更多合作项目落地见效，推动绿色丝绸之路建设走深走实。

2. 大力拓展绿色发展合作空间

中国与共建国家在绿色发展领域具有巨大合作潜力和广阔合作空间。中国愿与共建国家共同探讨，围绕绿色基建、绿色能源、绿色交通、绿色金融等重点领域，依托“一带一路”绿色发展国际联盟、“一带一路”能源合作伙伴关系等对话平台，共谋绿色发展之道，共筑清洁美丽世界。同时，积极培育绿色发展合作新增长点，不断拓展绿色科技、绿色产业、绿色贸易等领域合作，加强环境政策、规划、标准对接，提供更多“小而美”、可复制的绿色解决方案。进一步发挥“一带一路”生态环保大数据服务平台、一带一路环境技术交流与转移中心（深圳）等技术平台作用，促进先进绿色适用技术转移转化，推动绿色科技合作网络、绿色技术创新创业基地和国际环保产业园建设。

3. 持续强化绿色发展能力建设

能力建设是绿色发展的支撑保障，也是提升共建国家绿色发展水平的内在基础。中国愿同共建国家一道，坚持“授人以渔”，紧密围绕共建国家绿色增长和低碳发展现实需求，共同加强绿色发展能力建设，帮助共建国家提高绿色发展内生动力。聚焦环境管理、生物多样性和生态保护、应对气候变化、绿色金融、绿色交通与可持续基础设施建设等优先领域，更有针对性地开展绿色丝路使者计划、共建“一带一路”绿色发展专题培训、可持续投资能力建设联盟等能力建设活动，促进环境管理人员和专业技术人才培养交流。加大对共建“一带一路”绿色发展人才培养支持力度，为共建国家培养一批高素质、复合型、国际化生态环保人才，构建绿色发展智力支撑体系。

（二）携手加强绿色发展伙伴关系

1. 进一步深化多边合作

加强绿色丝绸之路建设与南南合作，与共建国家绿色发展战略、地区合作倡议对接，最大限度地凝聚各方共识，分享绿色发展理念与实践，携手推动全球环境治理体系朝向更加公平正义、包容并蓄的方向发展。依托《联合国气候变化框架公约》《生物多样性公约》等多边议程，实现经济发展和生态环境保护、应对气候变化协同增效。发挥中国—东盟环境保护合作中心、澜沧江—湄公河环境合作中心、中国—上海合作组织环境保护合作中心、中非环境合作中心等政府间合作平台作用，不断完善“一带一路”生态环境合作境外支点网络。

2. 进一步密切双边合作

绿色丝绸之路建设需要统筹考虑共建国家后疫情时代经济韧性发展与生态保护的现实需求。中国愿与共建国家深化政府间合作，对接绿色发展战略、政策、法律法规和标准，不断提升合作层次和水平，探索更加高效的合作机制。加强绿色发展重点问题研究及双边对话与沟通，推进环境管理制度、环保产业和技术等交流合作，不断拓展合作领域，丰富合作内容，提升合作水平，确保合作效果。立足现有中国与共建国家、国际组织签署的“一带一路”合作文件、双边合作战略协议等，持续推进现有双边环境合作框架下的务实合作，聚焦共建国家生态环保合作实际需求，打造双边绿色低碳示范项目，探索将最佳实践复制推广到更多共建国家。

3. 进一步促进民间交流

民心相通是国际合作的重要基础，绿色丝绸之路建设离不开绿色民心相通。中国愿同共建国家加强绿色发展领域的非政府组织交流合作，鼓励企业实施更多绿色低碳合作项目，增进共建国家政府、企业和公众的绿色共识。完善“一带一路”绿色发展民间交流机制建设，架起促进中外绿色人文交流、绿色民心相通的桥梁，增强共建国家民众的理解与认同，营造更加开放包容的“一带一路”绿色低碳合作氛围，为推进绿色丝绸之路建设不断向纵深方向发展厚植民意根基。

（三）携手共建人与自然生命共同体

1. 协同推进生态保护与污染治理

推进生态保护和污染治理是提升共建“一带一路”绿色化水平、实现可持续发展的基本前提。中国愿与共建国家加强战略对接和政策沟通，完善生态环境治理制度体系，提升生态环境管理水平。加强大气、水、土壤污染防治等领域合作，持续改善生态环境质量，推动污染源头治理，强化生态系统保护修复，协同推进降碳、减污、扩绿、增长。中国将继续引导中国企业提升生态环境管理能力，指导企业严格遵守东道国生态环保法律法规和规则标准，共同维护区域生态环境安全。

2. 协同推进生物多样性保护

生物多样性是人类赖以生存和发展的基础，是地球生命共同体的血脉和根基。中国愿与共建国家一道，切实履行《生物多样性公约》，积极承担与发展水平相称的国际责任，依托多双边平台机制不断深化生物多样性领域交流合作，推动实施“昆明—蒙特利尔全球生物多样性框架”，凝聚生物多样性保护治理合力，开启更加公正合理、各尽所能的全球生物多样性治理新进程。中国将发挥好昆明生物多样性基金作用，支持共建国家生物多样性相关战略规划制定与修订、生物多样性相关能力建设提升、信息交流、资金项目对接等生物多样性保护事业，推动实现人与自然和谐共生的美好愿景。

3. 协同推进应对气候变化行动

气候变化是全人类面临的共同挑战。中国愿与共建国家积极寻求应对气候变化“最大公约数”，在绿色丝绸之路建设框架下，为共建国家提供力所能及的资金、技术和能力支持，为应对气候变化、落实可持续发展目标提供解决方案。中国将与共建国家持续推动《联合国气候变化框架公约》及其《巴黎协定》全面平衡有效持续实施，分享彼此最佳实践和应对气候变化政策措施，强化对话交流与务实行动，推动构建公平合理、合作共赢的全球气候治理体系。中国将继续实施“一带一路”应对气候变化南南合作计划，推进低碳示范区建设，实施减缓、适应气候变化项目，提供应对气候变化相关援助，与共建国家共同提高应对气候变化能力，合力保护人类共同的地球家园。

附件

绿色丝绸之路发展大事记

2015 年

3 月 28 日，国家发展改革委、外交部、商务部联合发布《推动共建丝绸之路经济带和 21 世纪海上丝绸之路的愿景与行动》，提出“基础设施互联互通是‘一带一路’建设的优先领域……强化基础设施绿色低碳化建设和运营管理，在建设中充分考虑气候变化影响”；“在投资贸易中突出生态文明理念，加强生态环境、生物多样性和应对气候变化合作，共建绿色丝绸之路”。

9 月，中国宣布设立 200 亿元人民币的“中国气候变化南南合作基金”，支持其他发展中国家应对气候变化，并宣布于 2016 年启动“十百千”项目，即在发展中国家开展 10 个低碳示范区、100 个减缓和适应气候变化项目及 1 000 个应对气候变化培训名额的合作项目。

11 月 6 日，2015 年国际能源变革论坛期间，发布《苏州宣言》，指出能源转型的重要领域和方向。

2016 年

6 月 17 日，中国国家林业局与联合国防治荒漠化公约秘书处共同发布《“一带一路”防治荒漠化共同行动倡议》。

6 月 22 日，习近平主席在乌兹别克斯坦最高会议立法院的演讲中提出，要着力深化环保合作，践行绿色发展理念，加大生态环境保护力度，携手打造“绿色丝绸之路”。

7 月 26 日，习近平主席向 2016“一带一路”媒体合作论坛致贺信中提出，

“中国愿同沿线国家一道……携手打造‘绿色丝绸之路’、‘健康丝绸之路’、‘智力丝绸之路’、‘和平丝绸之路’，造福沿线国家和人民”。

10月30日，2016年国际能源变革论坛期间发布的《苏州共识》指出推动能源变革是化解全球能源资源和环境约束、实现可持续发展的重要途径。

12月11日，一带一路环境技术交流与转移中心（深圳）正式揭牌。该中心由环境保护部（现生态环境部）与深圳市政府共同建设，是中国与共建国家开展环保产业国际合作的实体平台。

12月11日，多家中国企业共同发起《履行企业环境责任 共建绿色“一带一路”倡议》，提出将在对外投资和国际产能合作中遵守环境法规、加强环境管理，助力绿色丝绸之路建设。

12月，环境保护部（现生态环境部）与联合国环境规划署签署《关于建设绿色“一带一路”的谅解备忘录（2019—2022年）》。

2017年

4月26日，环境保护部（现生态环境部）、外交部、国家发展改革委、商务部联合发布《关于推进绿色“一带一路”建设的指导意见》。

5月12日，环境保护部（现生态环境部）发布《“一带一路”生态环境保护合作规划》。

5月12日，国家发展改革委、国家能源局联合发布《推动丝绸之路经济带和21世纪海上丝绸之路能源合作愿景与行动》。

5月14日，习近平主席在首届“一带一路”国际合作高峰论坛开幕式上提出，“我们将设立生态环保大数据服务平台，倡议建立‘一带一路’绿色发展国际联盟，并为相关国家应对气候变化提供援助”。

5月15日，首届“一带一路”国际合作高峰论坛成果清单发布，包含《关于推进绿色“一带一路”建设的指导意见》《“一带一路”生态环境保护合作规划》、建设“一带一路”生态环保大数据服务平台、建立“一带一路”绿色发展国际联盟等成果。

6月14日，“一带一路”绿色供应链合作平台正式上线。

11月28日，澜沧江—湄公河环境合作中心在北京成立。

2018年

3月1日，国家林业局（现国家林业和草原局）印发《“一带一路”生态互联互惠科技创新行动方案》，提出创建“一带一路”生态互联互惠林业科技协同创新中心，促进绿色科技合作。

10月18日，在“一带一路”能源部长会议和国际能源变革论坛期间，中国与17个国家共同发布《建立“一带一路”能源合作伙伴关系联合宣言》，会议同时发布《中国可再生能源展望》《中国可再生能源发展报告》《中国城市能源报告》等成果。

11月30日，中国金融学会绿色金融专业委员会与伦敦金融城联合发起《“一带一路”绿色投资原则》。

2019年

4月16日，中国工商银行发行首笔“一带一路”银行间常态化合作机制绿色债券。

4月25日，中外合作伙伴在第二届“一带一路”国际合作高峰论坛之绿色之路分论坛上共同启动“一带一路”绿色发展国际联盟，正式发布“一带一路”生态环保大数据服务平台。

4月25日，在第二届“一带一路”国际合作高峰论坛期间，中国与29个国家在北京共同发起成立“一带一路”能源合作伙伴关系，并对外发布《“一带一路”能源合作伙伴关系合作原则与务实行动》。

4 月 25 日，在第二届“一带一路”国际合作高峰论坛期间，中国国家发展改革委与联合国人居署、世界卫生组织和能源基金会共同发起成立“一带一路”可持续城市联盟。

4 月 26 日，习近平主席在第二届“一带一路”国际合作高峰论坛开幕式上强调，“我们要坚持开放、绿色、廉洁理念……把绿色作为底色，推动绿色基础设施建设、绿色投资、绿色金融，保护好我们赖以生存的共同家园……我们同各方共建‘一带一路’可持续城市联盟、绿色发展国际联盟……我们启动共建‘一带一路’生态环保大数据服务平台，将继续实施绿色丝路使者计划，并同有关国家一道，实施‘一带一路’应对气候变化南南合作计划”。

4 月 27 日，第二届“一带一路”国际合作高峰论坛成果清单发布，包含“一带一路”绿色发展国际联盟、“一带一路”生态环保大数据服务平台、绿色丝路使者计划、“一带一路”应对气候变化南南合作计划、一带一路环境技术交流与转移中心、“一带一路”绿色照明行动倡议、“一带一路”绿色高效制冷行动倡议、“一带一路”能源合作伙伴关系、“一带一路”可持续城市联盟等。

11 月 19 日，中国生态环境部与柬埔寨环境部签署《关于合作建设低碳示范区的谅解备忘录》，支持柬埔寨提高应对气候变化能力。

12 月 3 日，首届“一带一路”绿色创新大会在深圳举行。

2020 年

7 月 16 日，中国生态环境部与老挝环境部签署《关于合作建设万象赛色塔低碳示范区的谅解备忘录》，支持老挝提高应对气候变化能力。

11 月 24 日，中非环境合作中心在北京成立。

12 月 1 日，“一带一路”绿色发展国际研究院在北京成立。

12 月 2 日，中国气候变化南南合作项目——援助埃塞俄比亚卫星交付仪式在北京举办，该卫星是中国第一颗援外卫星和同非洲合作的第一颗遥感卫星。

2021年

1月9日，中国生态环境部与塞舌尔共和国农业、气候变化和环境部签署《关于应对气候变化南南合作低碳示范区建设实施的谅解备忘录》，支持塞舌尔提高应对气候变化能力。

6月23日，29国在“一带一路”亚太区域国际合作高级别会议期间，共同发起“一带一路”绿色发展伙伴关系倡议。

6月28日，2021年国际能源变革对话活动期间，49家企业联合发布《加快能源低碳发展苏州倡议》。

7月9日，商务部、生态环境部联合发布《对外投资合作绿色发展工作指引》。

9月21日，习近平主席在第七十六届联合国大会一般性辩论上发表讲话，强调“我们必须复苏经济，推动实现更加强劲、绿色、健康的全球发展……中国将大力支持发展中国家能源绿色低碳发展，不再新建境外煤电项目”。

10月18至19日，第二届“一带一路”能源部长会议期间，“一带一路”能源合作伙伴关系发布《“一带一路”绿色能源合作青岛倡议》《“一带一路”能源合作伙伴关系章程》以及能源国际合作最佳实践等多项成果。

10月24日，国务院印发《2030年前碳达峰行动方案》，强调推进绿色丝绸之路建设，加强与共建“一带一路”国家的绿色基建、绿色能源、绿色金融等领域合作，发挥“一带一路”绿色发展国际联盟等合作平台作用。

10月28日，中国政府向《联合国气候变化框架公约》秘书处正式提交《中国落实国家自主贡献成效和新目标新举措》，指出“中国坚持绿色低碳发展，打造‘绿色丝绸之路’，积极推动实施‘一带一路’应对气候变化南南合作计划……充分发挥‘一带一路’绿色发展国际联盟等多双边合作平台的作用，推动全球气候变化治理及绿色转型专题伙伴关系建设，为应对气候变化国际合作汇聚更多力量”。

11 月 10 日，第二届中国—中东欧国家企业能源合作论坛期间，《中国—中东欧企业深化清洁能源合作宁波宣言》发布。

2022 年

1 月 5 日，生态环境部、商务部联合发布《对外投资合作建设项目生态环境保护指南》。

1 月 20 日，生态环境部与中国出口信用保险公司签署《协同推进绿色丝绸之路建设战略合作协议》，共同服务支持绿色丝绸之路建设。

3 月 16 日，国家发展改革委、外交部、生态环境部、商务部联合印发《关于推进共建“一带一路”绿色发展的意见》。

7 月 12 日，“一带一路”绿色发展国际联盟和“一带一路”媒体传播联盟共同启动“绿色丝路行”国际传播活动。

11 月 9 日，中国生态环境部与柬埔寨环境部签署《关于开展应对气候变化南南合作低碳示范区第二批物质援助项目的合作谅解备忘录》，继续支持柬埔寨提高应对气候变化能力。

11 月 24 日，中亚区域经济合作第 21 次部长级会议召开，会议以“重振区域合作，实现绿色、可持续及包容复苏”为主题，中国及 9 个中西亚区域国家通过了《中亚区域经济合作绿色、可持续、包容性疫后复苏框架》《中亚区域经济合作绿色能源联盟》等成果文件。

12 月 15 日，习近平主席在《生物多样性公约》第十五次缔约方大会（COP15）第二阶段高级别会议开幕式致辞中指出，要“依托‘一带一路’绿色发展国际联盟，发挥好昆明生物多样性基金作用，向发展中国家提供力所能及的支持和帮助，推动全球生物多样性治理迈上新台阶”。

2023 年

1 月 19 日，国务院新闻办公室发布《新时代的中国绿色发展》白皮书，指出中国始终致力于推进共建“一带一路”绿色发展，让绿色切实成为共建“一带一路”的底色。

4 月 12 日，“一带一路”全民早期预警高层论坛期间，中国生态环境部、中国气象局与世界气象组织签署关于支持联合国全民早期预警倡议的三方合作协议。

5 月 10 日，“一带一路”绿色发展国际联盟召开首次会员大会，成为首个在绿色丝绸之路框架下成立的国际性环保社会团体。

5 月 19 日，习近平主席主持首届中国－中亚峰会并发表主旨讲话，指出“要继续在共建‘一带一路’合作方面走在前列，推动落实全球发展倡议，充分释放传统合作潜力，打造减贫、绿色低碳等新增长点，携手建设一个合作共赢、相互成就的共同体”。

5 月 24 日，第三届“一带一路”能源合作伙伴关系论坛期间，《绿色金融支持“一带一路”能源转型倡议》发布。

8 月 3 日，中国生态环境部（国家核安全局）与巴基斯坦核监管局签署《关于成立中国—巴基斯坦核安全合作中心的谅解备忘录》。

9 月 6 日，2023 年国际能源变革论坛期间，中国发布《全球清洁能源合作伙伴关系倡议》，43 家中外能源企业和研究机构共同发布《发起筹备国际能源变革联盟倡议》，首次建立国家能源变革量化评估体系并发布《能源变革指数蓝皮书》。

9 月，中国交通运输部与联合国签署推动可持续交通促进可持续发展目标的合作谅解备忘录。

9月11日，中国银行发行全球首批共建“一带一路”主题绿色债券。

10月18日，习近平主席在第三届“一带一路”国际合作高峰论坛开幕式主旨演讲中提出，中方将持续深化绿色基建、绿色能源、绿色交通等领域合作，加大对“一带一路”绿色发展国际联盟的支持，继续举办“一带一路”绿色创新大会，建设光伏产业对话交流机制和绿色低碳专家网络。落实《“一带一路”绿色投资原则》，到2030年为伙伴国开展10万人次培训。

10月18日，生态环境部、国家发展改革委共同主办第三届“一带一路”国际合作高峰论坛绿色发展高级别论坛。论坛上发布了《“一带一路”绿色发展北京倡议》、绿色发展投融资合作伙伴关系与中亚区域绿色科技发展行动计划三项成果。

10月18日，商务部主办第三届“一带一路”国际合作高峰论坛贸易畅通专题论坛。专题论坛期间，中国与34个国家共同发布了《数字经济和绿色发展国际经贸合作框架倡议》。

Report on Development of the Green Silk Road

(2023)

Office of Leading Group for Promoting the Belt and Road Initiative

Ministry of Ecology and Environment of the People's Republic of China

China Environment Publishing Group · Beijing

图书在版编目（CIP）数据

绿色丝绸之路发展报告．2023：汉文、英文 / 推进“一带一路”建设工作领导小组办公室，生态环境部编著．-- 北京：中国环境出版集团，2023.9
ISBN 978-7-5111-5633-4

Ⅰ．①绿… Ⅱ．①推… ②生… Ⅲ．①丝绸之路－生态环境保护－研究报告－2023 Ⅳ．①X321.2

中国国家版本馆 CIP 数据核字（2023）第 184273 号

出 版 人 武德凯
责任编辑 黄　颖　梅　霞
装帧设计 宋　瑞

出版发行 中国环境出版集团
（100062　北京市东城区广渠门内大街 16 号）
网　　址：http://www.cesp.com.cn
电子邮箱：bjgl@cesp.com.cn
联系电话：010-67112765（编辑管理部）
发行热线：010-67125803，010-67113405（传真）
印　　刷 玖龙（天津）印刷有限公司
经　　销 各地新华书店
版　　次 2023 年 9 月第 1 版
印　　次 2023 年 9 月第 1 次印刷
开　　本 787×1092　1/16
印　　张 7.25
字　　数 100 千字
定　　价 80.00 元（全 2 册）

Contents

Contents

Preface

Currently, the world is witnessing unprecedented changes in a century. With the deepening trends of multi-polarization, economic globalization, social digitalization and cultural diversification across the globe, the destiny of people in various countries have never been so closely connected as they are today. Faced with the epochal challenges of the times, in the autumn of 2013, Chinese President Xi Jinping successively proposed the major initiative of jointly building the "Silk Road Economic Belt" and the "21st Century Maritime Silk Road" (hereinafter referred to as the BRI) during his visits to Kazakhstan and Indonesia. In March 2015, China published the *Vision and Actions on Jointly Building the Silk Road Economic Belt and the 21st Century Maritime Silk Road*, clearly proposing to jointly build a Green Silk Road. In May 2017, at the First Belt and Road Forum for International Cooperation, Chinese President Xi Jinping proposed to practice the new concept of green development, advocate green, low-carbon, circular, and sustainable production and lifestyle, strengthen cooperation in ecological and environmental protection, boost the construction of ecological civilization and achieve the 2030 Sustainable Development Goals. In April 2019, President Xi Jinping further emphasized at the Second Belt and Road Forum for International Cooperation that in making the BRI a success, "green" should be made a defining feature through launching green infrastructure projects, making green investments, and providing green financing to protect the Earth which we all call home. In October

2023, President Xi Jinping announced eight steps to support the high-quality Belt and Road Cooperation at the Third Belt and Road Forum for International Cooperation. As one of the eight major steps, promoting green development has pointed out new direction for BRI green development, opening up new prospect and inject new impetus for the future.

To make the Silk Road a green one is the intrinsic requirement for practicing the green development concept and advancing ecological progress. It is a major step to actively respond to climate change and maintain global ecological security. It is also a vehicle to achieve the high-quality development of the BRI and build a community of life for humanity and nature. Over recent years, China has continuously improved the top-level design of the Green Silk Road, launched a series of green action initiatives, and implemented a number of green development projects, thus taking the development of the Green Silk Road to a higher level.

— Striving for a higher level for a broader view, letting the top-level design lead the way. Under the unified deployment of the leading group for promoting the BRI, all participating organizations have worked together to establish and improve the policy system for developing the Green Silk Road. Since 2017, the Chinese government has published a series of documents, including *Guidance on Promoting Green Belt and Road*, *Belt and Road Ecological and Environmental Cooperation Plan*, and *Opinions on Jointly Promoting Green Development of the Belt and Road* to specify overall requirements, deploy key tasks and working arrangements. Besides, the green development of the BRI has also been highlighted as a priority in such key documents as the *China's Outline of the 14th Five-Year Plan (2021-2025) for National Economic and Social Development and the Long-Range Objectives Through the Year 2035*, *Working Guidance for Carbon Dioxide Peaking and Carbon Neutrality* *in Full and Faithful Implementation of the New Development Philosophy of the CPC Central Committee and State Council*, and *Notice from the State Council on Issuing the Action Plan for Carbon Dioxide Peaking Before 2030*.

— Bringing countries together by fulfilling our responsibilities. Among the over 200 cooperation documents signed between China and over 150 countries and 30 international organizations on the joint construction under BRI, Green Silk Road has always been a key component. The BRI International Green Development Coalition (BRIGC) has brought together more than 150 Chinese and foreign partners and held over 70 thematic events on the construction of the Green Silk Road of which the influence spreads over to around 160 countries and regions. *The Green Investment Principles for the Belt and Road Development* has included 46 signatory institutions and 17 supporting institutions from 17 countries and regions holding or managing assets of over USD 41 trillion. The Chinese government has organized more than 350 training sessions and academic degree projects on protecting the environment and responding to climate change, which helped train more than 11,000 professionals from relevant countries. Among these projects, the Green Silk Road Envoys Program and the Belt and Road South-South Cooperation Initiative on Climate Change have trained more than 3,000 officials and technicians working on the environment and climate across more than 120 developing countries.

— Taking down-to-earth steps, seeking common development through concrete measures and cooperation. Currently, green and low-carbon transformation has become a mega trend in the world. In recent years, major steps have been taken to build a beautiful China, and green development has become a prominent feature of Chinese modernization, which has benefited not only China but also the world. According to data released by the National Energy Administration, in 2022, China's exports of wind power and photovoltaic products reduced carbon dioxide emissions by approximately 573 million tons for other countries, with a total emission reduction of 2.83 billion tons, accounting for approximately 41% of the global converted carbon emission reductions from renewable energy in the same period. Meanwhile, China has cooperated with more than 100 countries in solar power, wind power, hydropower, thermal energy and other fields, which has injected a strong impetus to accelerate the transformation of energy supply towards a more efficient, clean, and diversified direction for these countries.

Although the notion of BRI was proposed by China, the opportunities and achievements are shared by the world. Green, low-carbon, and sustainable development concerns the common destiny of all mankind and the interests and well-being of current and future generations, thus requiring the joint efforts of the international community. Looking into the future, the Green Silk Road conforms to the global trend of green and low-carbon development, encapsulates the new expectations of people around the world for a beautiful environment, and provides experience, paths and solutions for the BRI participating countries to implement the United Nations 2030 Agenda for Sustainable Development. Therefore, it will show greater vitality in the new environment. Sharing the same wishes and the same goals, China will join hands with other BRI participating countries to build a green foundation for high-quality development, adhere to the concept of harmonious coexistence between man and nature, and follow the principle of discussion and collaboration. Aiming at high standards, benefiting people's livelihood, and sustainability, China will continue to improve the green development cooperation, build a closer green development partnership, jointly build a community of life for humanity and nature, and write a new chapter of high-quality BRI cooperation together.

I. Development Achievements

With the joint efforts of China and BRI participating countries, the construction of Green Silk Road has yielded fruitful development outcome. The "hardware connectivity" of green infrastructure, "software connectivity" of green development rules and standards, and "people-to-people ties" through green exchanges and cooperation have all been strengthened, carving a new path for developing the economy while protecting the environment at the same time, making green a feature of the high-quality Belt and Road cooperation and infusing great power into the green, low-carbon transformation and sustainable development of the BRI participating countries.

1. Building Consensus on Green Development in International Cooperation

1) Deepening coordination on green development policies

— Promoting coordination of green development strategies. China has signed over 200 BRI cooperation documents with more than 150 countries and 30 international organizations, of which Green Silk Road is an important component. China launched the Initiative for Belt and Road Partnership on Green Development together with 28 countries, jointly launched the *Beijing Initiative for Belt and Road Green Development* with 21 government and environmental authorities of BRI participating countries and international organizations, launched the *Initiative* *on International Trade and Economic Cooperation Framework for Digital Economy and Green Development* with 34 partner countries, and signed the *Memorandum of Understanding on Building a Green Belt and Road (2017-2022)* with the United Nations Environment Programme. The relevant consensus has been incorporated into important bilateral documents and outcome documents of cooperation mechanisms between China and ASEAN, five Central Asian countries, Central and Eastern European countries, Pacific Island countries, as well as China-Arab States Cooperation Forum, China-Africa Cooperation Forum, and China-CELAC Forum. Cooperation documents concerning environmental protection, sustainable development, and response to climate change have been signed between China and environment and climate related authorities of Iran, Belarus, Pakistan and other countries. China has signed memorandums of understanding on investment and cooperation in green development with the governments of Russia, Hungary, Kyrgyzstan, Argentina, Serbia, Singapore, Indonesia and Papua New Guinea to continuously expand cooperation in clean energy, energy conservation, emission reduction, and green technology innovation.

— **Establishing regional cooperation mechanism in key areas.** China has carried out green exchanges and cooperation with BRI participating countries under the UN and G20 frameworks, and carried out dialogues and communication through multilateral and bilateral cooperation mechanisms such as the UN Environment Assembly, the G20 Environment Ministers' Meeting, and the China-ASEAN Dialogue on Climate Change and Environment. Several cooperation centers have been established including the China Center for SCO Environmental Cooperation, Lancang-Mekong Environmental Cooperation Center and China-Africa Environmental Cooperation Center, which have become important collaborative platforms for relevant government departments, international organizations, research institutions and enterprises.

2) Establishing a platform for green exchanges and cooperation

— Establishing a platform for policy dialogue and information sharing. During the Second BRI Forum for International Cooperation, China and partner countries jointly kicked off the BRI International Green Development Coalition (BRIGC) and the BRI Environmental Big Data Platform. As of August 2023, more than 150 partners from 43 countries, including the environmental authorities of 26 BRI participating countries, have joined the BRIGC, making it the first international social organization under the Green Silk Road framework. So far, it has held over 70 seminars, including the BRI Green Innovation Conference, released more than 20 policy research reports such as the *Green Development Guidance for BRI Projects*, initiated the Green Investment and Finance Partnership (GIFP), and launched the international communication program named Vista of the Green Silk Road with the Belt and Road Media Cooperation Union. The BRI Environmental Big Data Platform has gathered basic environmental information, environmental laws, regulations, and standards from over 60 BRI participating countries, and over 200 metrics data disclosed by more than 30 international authoritative platforms. It has completed the integration of metrics data for specific projects, including atmospheric environment, biological diversity, water resources and water environment, etc.

— Building a professional platform for practical cooperation. In April 2019, China jointly initiated the Belt and Road Energy Partnership with relevant BRI participating countries in the energy field, marking the first international energy cooperation platform initiated by China. As of November 2023, the member states have reached 33 countries. So far, two Belt and Road Energy Ministerial Conferences and three Partnership Forums were hosted and yielded various outcomes, including the *Qingdao Initiative for Belt and Road Green Energy Cooperation*, *the Belt and Road Energy Partnership Charter*, and *the Best Practice Cases of Belt and Road Energy Cooperation*, etc. In the field of urban sustainable development, the Belt and Road Sustainable Cities Alliance (BRSCA) has been launched by the National Development and Reform Commission of China, in partnership with the United Nations Human Settlements Programme, the World Health Organization and the Energy Foundation. The BRSCA has hosted four International High-Level forums on Sustainable Urban Development in a row, issued the *Green Urbanization Consensus* and formed a series of research reports to share China's urban sustainable development experience and cases with 27 BRI participating countries and 190 trainees. In the field of green finance, the Green Finance Committee of China Society for Finance and Banking (GFC) and the City of London jointly launched the Green Investment Principles for the Belt and Road in November 2018, which now includes 46 signatory institutions and 17 supporting institutions from 17 countries and regions.

3) Improving capabilities of green development

China actively helps the BRI participating countries improve their capabilities and strengthen interaction and exchanges between environmental officials and professional technical personnel, so as to provide talent support for green development. The implementation of the Green Silk Road Envoys Program and the Belt and Road South-South Cooperation Initiative on Climate Change, organized nearly 100 training activities on environment and climate change, provided training sessions for over 3,000 professionals from more than 120 developing countries, launched the BRI Green Talent Program, and set up academic degree programs on environment management and climate change, which helped train more than 11,000 professionals for relevant countries. Such efforts are widely recognized by participating countries, international organizations and various social sectors. Relevant Chinese institutions have carried out targeted capacity development projects based on the actual needs of different regions. They have implemented the China-Africa Green Envoys Program, organized training programs to support the construction of Africa's Great Green Wall Initiative. The International Training Program on Rainwater Collection and Utilization Technology has been held for many years, and Chinese experts went to Jordan, Saudi Arabia, Nigeria, South Africa and other countries to train local professionals. All of these efforts helped China win high recognition from international organizations such as the United Nations and relevant industrial associations.

2. Promoting the Construction of Green Infrastructure

1) Building green infrastructure

In July 2021, to promote green concepts in overseas investment projects, the Chinese government issued the *Guidelines for Green Development of Outbound Investment and Cooperation*, proposing to build green infrastructure, establish green overseas economic and trade cooperation zones, support green technology innovation, promote green transition of enterprises, and encourage and guide companies to raise their awareness of green development. In January 2022, the Chinese government issued the *Guidelines for Ecological and Environmental Protection in Overseas Investment and Cooperation Construction Projects*, which focuses on the ecological and environmental management in the entire life cycle of overseas investment cooperation and construction projects, and further standardizes the environmental protection behavior of enterprises in overseas investment and cooperation. As a result, Chinese enterprises have actively practiced green and low-carbon concepts throughout the project construction, showcasing with their actions that green is the defining feature of BRI, and making positive contributions to the economic and social development as well as ecological and environmental protection of the countries where the projects are located.

Column I: Supporting BRI Participating Countries in building Green Infrastructure

Mombasa-Nairobi Railway

The Mombasa-Nairobi Railway is a BRI flagship project jointly built by China and Kenya. It was constructed by Chinese enterprises and opened to operation in May 2017. With a total length of 471 kilometers, the Mombasa-Nairobi Railway connects Nairobi, the capital of Kenya, and Mombasa, the largest port in East Africa. It is the first overseas railway constructed entirely in accordance with "Chinese standards", leaving a green mark of BRI in Africa. The railway design gives full consideration to the living habits and migration paths of wild animals. A total of 14 large wildlife passages and 79 bridges have been designed along the entire line, with heights of over 6.5 meters, ensuring the smooth passage of large animals such as giraffes. At the same time, more than 100 culverts have been set up to provide drinking water sources and migration channels for animals. In addition, the project has specially drew up the red line for ecological protection to protect the local mangrove wetland park. The Mombasa-Nairobi Railway operates an average of 9 pairs (18 trains) of freight trains per day, using far less diesel fuel than road transport, reducing 75% of greenhouse gas emissions.

2.Ghana Tema New Container Terminal Project

The Ghana Tema New Container Terminal Project located in the Gulf of Guinea in western Africa, is the largest water engineering project in Ghana in recent years. The project was constructed by Chinese companies and was completed in June 2020. Ghana is an important breeding base for sea turtles. Before the construction began, an environmental baseline survey was conducted with an environmental impact assessment. An environmental management plan during the construction period was formulated to ensure that health, safety, environment and social responsibility meet the requirements of the World Bank International Finance Corporation and Ghana Environmental Protection Agency. The project has specifically set up a Sea Turtle Hatchery to strengthen the in-situ protection of sea turtles within the project area. By working together with Ghana Wildlife Society, sea turtle hatching experts are hired as advisors. As of June 2021, a total of more than 17,000 sea turtles have been hatched and returned to nature with a hatching rate at 78.8% (that figure under natural conditions is about 10%-20%). All these measures have played an important role in protecting species and maintaining ecological balance.

2) Building green and low-carbon industrial parks

In the process of jointly building the economic and trade cooperation parks, China adheres to high-standard in planning development goals and seting up a management systems, actively practicing the concept of green development in environmental infrastructure construction, promoting the centralized treatment and recycling of wastewater and garbage in industrial parks, and striving to reduce carbon emissions. It has strengthened the linkage between scientific and technological innovation and industrial innovation, promoted relevant enterprises to adopt international advanced standards by formulating relevant quality standards and industry technical specifications. Meanwhile, with the introduction of green industry and projects, such as photovoltaic power generation, new energy vehicles, clean energy equipment, waste incineration power generation, waste recycling, etc., a circular route among relevant industries within the park has been established for the recycling of resources, which constantly improves green, low-carbon and circular development of the park.

Column II: Supporting BRI Participating Countries in Building Green and Low-Carbon Industrial Parks

1. Thai-Chinese Rayong Industrial Zone

The Thai-Chinese Rayong Industrial Zone, jointly built by China and Thailand, explicitly requires companies to install wastewater treatment facilities and set discharging standards. The project is also asked to reduce its impact on the local community and landscape, and establish complaints and response mechanisms. As an exemplary pilot of green industry, the Industrial Zone takes environmental standards as a prerequisite for entry and adds toxic, harmful, and high-polluting projects to the exclusion list so as to prevent the entry of the projects that may harm the environment from the root. Enterprises in biotechnology, new energy, new materials and other industries are encouraged to settle in, and they are reguired to carry out plant construction accorrding to ISO14001 environmental protection standards and accept third-party wastewater treatment supervision and evaluation.

2. Indonesia Tsingshan Industrial Park

The Tsingshan Industrial Park, jointly built by China and Indonesia, is designed to be a clean energy base with solar, wind power, etc. to support the production of batteries and other components. Taking advantage of the rich local nickel reserves, the park takes the integrated production of "nickel-iron + stainless steel" as the leading industry to coordinate industrial development and environmental protection. With advanced technologies, carbon dioxide emission per ton of nickel production is reduced by more than 60%. The Tsingshan Industrial Park has also taken actions, such as marine garbage management, coral transplantation and mangrove protection, to ensure the continuing increase of local tropical fishes and coral reefs, and reduce the impact of the project on the local ecological environment and biodiversity to the largest extent. The project was honored with the "Outstanding Industrial Park Award" by the Indonesian government, has become a demonstration project of industrial cooperation between the two countries.

3) Building environmental infrastructure

Wastewater and solid waste treatment is an arduous environmental challenge faced by developing countries in the process of rapid industrialization and urbanization. In line with the actual needs of the locality, under the principles of green, low-carbon, intensive, efficient and circular development, Chinese enterprises help BRI participating countries improve their environmental infrastructure, constantly improve the green and intelligent level of environmental infrastructure, properly utilize local resources and treat pollution, largely improve the living conditions for the locals and achieve a coordinated development of economy, society and environment by jointly building centralized wastewater treatment facilities and solid waste disposal facilities, and adopting advanced technology.

Column III: Supporting BRI Participating Countries in Building Environmental Infrastructure

1. Sewage Treatment Facilities

The Central Wastewater Treatment Plant of Ulaanbaatar, jointly built by China and Mongolia, is expected to be the largest, most modern and sophisticated sewage treatment plant in Mongolia upon completion. The project will benefit 44% of Mongolia's population, effectively improve the previously inadequate capacity in sewage treatment of Ulaanbaatar, and ensure the safety of water for people to use. The Dasherkandi Sewage Treatment Plant jointly built by China and Bangladesh is the first modern and large-scale sewage treatment plant in the country and the largest single sewage treatment plant in South Asia. It can process domestic sewage for nearly 5 million people each day in the urban area of Dhaka, greatly improving the water environment quality of Dhaka. The Kokopo sewerage treatment project, jointly built by China and Papua New Guinea, covers sewage pipelines, sewage treatment and road greening projects. Through oxidation, filtration, sterilization, disinfection, testing and other process to make sewage meets the discharge standards, significantly improving the local environment. In addition, in 2023, China actively promoted the exploration and approval of the sewage treatment plant project in the Syr Darya River of Uzbekistan.

2. Solid Waste Treatment Facilities

The Can Tho Waste-to-Energy project, invested, constructed and operated by a Chinese enterprise, is Vietnam's first modern and high-standard waste-to-energy project. Supported by China's advanced technology, this project has greatly reduced solid waste and improved waste-to-energy-related technology. It won the Can Tho City Excellent Enterprise Award in 2018. Another solid waste treatment project in Colombo, Sri Lanka's commercial capital, adopts advanced membrane technology. The treated landfill leachate can meet the country's Class I discharge standard and can be directly used as the irrigation water for green spaces in the Puttalam area, thus enabling the recycling and resource utilization of solid waste. The Bangkok Nong Khaem waste-to-energy project, jointly developed by Chinese and Thai companies, has been operating stably and up to relevant emission standards since it was connected to the local power grid in March 2016. It was also certified by the International Organization for Standardization (ISO) management system. The Chinese enterprise has been highly recognized by the local government for its outstanding performance in living up to its environmental and social responsibilities.

3. Deepening Exchanges and Cooperation in Green Energy

1) Jointly developing renewable energy projects

In recent years, China has achieved leapfrog development in renewable energy and has become a leader and contributor in the global renewable energy sector. According to the International Renewable Energy Agency, as of the end of 2022, the global installed capacity of renewable energy totaled 3.37 terawatts (TW). Almost half of the newly added capacity in 2022 was in Asia. Among them, China contributed the most, with 141 gigawatts (GW) of new installed capacity. According to data released by the National Energy Administration of China, in 2022, China's renewable energy power generation was equivalent to the reduction of approximately 2.26 billion tons of domestic carbon dioxide emissions. The exported wind power and photovoltaic products reduced carbon dioxide emissions by approximately 570 million tons for other countries. Consequently, the total emission reduction reached 2.83 billion tons, accounting for approximately 41% of the world's carbon emission reductions converted from renewable energy generation during the same period. Meanwhile, China has cooperated with more than 100 countries in solar power, wind power, hydropower, thermal energy and other fields, which has injected a strong impetus to accelerating the transformation of their energy supply towards a more efficient, clean and diversified direction. A large number of clean, efficient and high-quality renewable energy projects have been completed successively, including the Adama Wind Power Project in Ethiopia, the Reed Bridge Wind Power Project in Ecuador, the Cauchari Solar Power Project in Jujuy Province of Argentina, the Mozura Wind Park in Montenegro, and the Jiu'an Wind Power Project in Vietnam. These projects have effectively solved the problem of inaccessible and expensive electricity supply in local areas, greatly optimizing the energy structure, reducing pollutants and greenhouse gas emissions and promoting economic development locally.

Column IV: Supporting BRI Participating Countries in Implementing Projects on Renewable Energy

1. Pakistan Karot Hydropower Project

The Pakistan Karot Hydropower project is the first hydropower project under the China-Pakistan Economic Corridor and was officially put into operation in June 2022. With a total installed capacity of 720,000 kilowatts and an average annual generating capacity of 3.2 billion kilowatt-hours, the hydropower plant is expected to meet the electricity needs of about 5 million local people and save 1.4 million tons of standard coal every year. The construction abided by the Environmental and Social Framework and Standards set by the World Bank International Finance Corporation and fully adopted international standards in the environmental impact assessment. Shahid Khaqan Abbasi, the former Prime Minister of Pakistan, hailed the project as a model that benefits the country and the people. The project strictly implements the concept of green development. In-depth publicity and education programs on ecological environment protection were carried out in the affected area before the construction of the project, and regulations on protecting fish resources were specifically formulated to preserve the aquatic habitats downstream of the dam. After its completion, the project is expected to reduce carbon dioxide emissions by 3.5 million tons per year.

2. Montenegro Mozula Wind Park

The Mozula Wind Park is a benchmark renewable energy project in Montenegro and is expected to help the nation achieve its goal of "doubling the installed capacity of new energy". The project generates 110 million kilowatt-hours of clean electricity per year, accounting for about 5% of the total electricity generated in Montenegro, and can reduce greenhouse gas emissions by about 3,000 tons per year. The project has made great contributions to increasing the share of renewable energy and promoting emission reduction, bringing significant social benefits to Montenegro. After its completion, the project is expected to meet the electricity demand of 100,000 residents, greatly alleviating the problem of frequent power outages during the peak tourist season. The optimization of the energy environment has also attracted more foreign investment thus creating more jobs. The project adopts the advanced "smart wind power" technology to increase power generation efficiency by 15%-20%. With the smart wind farm software developed by the China, the operation cost can be reduced by 20%.

3. Cauchari Solar Power Station Project in Jujuy Province, Argentina

The Cauchari solar power station project in Jujuy Province, Argentina, is the largest installed capacity and the highest altitude photovoltaic power station project in South America. The total installed capacity is about 300 megawatts. The design annual average on-grid electricity is about 660 million degrees, can save about 202,000 tons of standard coal per year, can reduce 552,000 tons of carbon dioxide, 106.1 tons of sulfur dioxide, 118.7

tons of nitrogen oxides, and 21.2 tons of soot on an annual basis. The project was officially connected to the grid in September 2020, and put into operation in April 2021. In 2021, the project generated 711 million kilowatt-hours of electricity in accumulation, which has effectively alleviated the strained local power supply, promoted the development of local industries and tourism. It has played a positive role in improving the employment and promoting economic and social development in local area.

4. Wind Power Project in Jiu'an, Vietnam

Jiu'an Wind Power Project is located in Jiu'an of Gia Lai Province in Vietnam. It is the first local wind power project to be connected to the power grid, setting a new benchmark for wind power projects in Gia Lai Province. More than 90% of the equipment used in the project is made by China. With a total capacity of 46.2 megawatts, it can provide 110 million kilowatt-hours of clean electricity every year and reduce carbon dioxide emissions by about 90,000 tons, supporting Vietnam achieving the net-zero carbon emission goal by 2050. The project further optimizes the local energy mix and increases the share of renewable energy. The green electricity generated can meet the demands of approximately 50,000 local residents, effectively alleviating the strained local power supply. During the construction and operation of the project, it provided about 80 jobs for local people, thus playing a positive role in promoting local employment and economic development.

5.Dubai Solar Thermal and Photovoltaic Solar Power Plant, United Arab Emirates

The 950MW solar thermal and photovoltaic hybrid power generation project in Dubai, United Arab Emirates, consists of three trough power plants with an installed capacity of 200MW, one tower power plant with an installed capacity of 100MW, and one photovoltaic power plant with an installed capacity of 250MW. It is the largest solar thermal and photovoltaic hybrid power plant in the world to date, and also the largest solar thermal project with the largest installed capacity, investment scale and molten salt thermal storage in the world. After completion, the project will have the world's largest thermal storage capacity, providing clean electricity to over 270,000 households in Dubai and reducing carbon emissions by 1.4 million tons annually.

2) Promoting low-carbon energy transition and upgrading

— **Stopping building new overseas coal-fired power projects.** The proportion of fossil fuels in overseas energy investment and construction projects in which China participates has been declining since 2016. In 2020, the total share of investment and construction projects in renewable energy (hydropower, photovoltaic, wind energy and biomass energy) exceeded that in fossil fuels for the first time. On September 21, 2021, Chinese President Xi Jinping announced at the General Debate of the 76th Session of the UN General Assembly that China would vigorously support green and low-carbon energy development in developing countries and would no longer build new overseas coal-fired power projects. This is another contribution made by China in tackling climate change since Chinese President Xi Jinping's announcement that China would peak carbon emissions before 2030 and achieve carbon neutrality before 2060. It is also another positive contribution made by China to promote global green recovery, and show its firm commitment and practical actions to build a community of life for humanity and nature, which has been highly recognized and acclaimed by the international community.

— Promoting cooperation in energy conservation in key areas. Some BRI participating countries are now going through the stage of accelerated industrialization. The resource and energy consumption as well as pollution discharge of these countries are experiencing rapid growth with strong demands for energy-intensive products such as steel, cement and petrochemicals. Chinese enterprises have helped BRI participating countries adopt technologies to improve efficiency and reduce consumption, upgrade energy systems, and bring in big data analytics to reduce pollution and conserve water and electricity. The Shymkent Oil Refinery built in the 1970s is one of the three largest refineries in Kazakhstan. After it was upgraded to protect the environment by the Chinese enterprise, pollution discharge has been reduced by 90%, and about 1.5 million cubic meters of water is saved every year. The CISDI-Formosa Ha Tinh Steel Plant Project of Vietnam adopts a key technology cluster for extra-large blast furnaces featuring high-efficiency and low energy consumption, as well as the blast furnace intelligent production management system independently developed by China. It combines big data analytics with the actual production process, which have achieved green and energy-saving production of steel. The main technical indicators have reached the international advanced level.

3) Strengthening cooperation on energy technology and equipment

Technological innovation is a major breakthrough for green and low-carbon transformation. As the world's largest green energy market and equipment manufacturer, China has established a green energy industry system covering the entire industrial chain with strong capabilities in research and development, production, installation, operation and maintenance. The manufacturing scale of equipment for wind power and solar power generation ranks first in the world, and the photovoltaic modules, wind turbines, gearboxes and other key components produced by China account for 70% of the global market share. By strengthening outbound investment in green energy and promoting environmental technology assistance and cooperation, China has vigorously pushed its green energy equipment, technology and services to "go global" to support BRI participating countries to achieve the green energy transition. Jinko Solar invested to build a solar cell and module production plant in Penang, Malaysia in 2015. The plant now produces 7GW photovoltaic cells and 7GW modules on an annual basis. In addition to manufacturing facilities, Jinko Solar has also established the largest overseas product R&D center among Chinese photovoltaic companies in Malaysia for the development, experimentation and testing of photovoltaic modules, which has promoted the exchange of advanced photovoltaic products and technologies between China and Malaysia.

4. Fostering Green and Low-Carbon Transportation

1) Constructing green transportation infrastructure

In September 2023, the Ministry of Transport of China signed the *MOU on Promoting Sustainable Transport for the SDGs* with the United Nations. During the implementation of such transportation infrastructure projects as those on railways, highway, and airports in BRI participating countries, Chinese enterprises have integrated low-carbon and environmental protection concepts into the whole process from project planning, and construction to operation. In the survey and design stage, plans are made to avoid, decrease and even eliminate the impact on protected areas and ecologically sensitive and fragile areas. Technologies in energy conservation, noise reduction, environmental governance, intensive resource utilization and sharing are extensively adopted to make transportation infrastructure standardized and more intelligent. The construction of permanent facilities and make-shift facilities are combined, and the integration of construction and maintenance are promoted so as to reduce energy consumption and carbon emissions in the whole life cycle of projects. Taking the China-Laos Railway as an example, Chinese enterprises have implemented high-standard ecological protection measures in the project. From survey, route selection to construction, all possible environmental impacts of the project have been fully considered. It skillfully integrates railway with nature, combines the ethnic characteristics and cultural heritage protection along the route with a green transportation concept. The project fully considered the opinions of local communities, experts and environmentalists during the construction process, winning recognition from the local people for the greening performance of Chinese enterprises.

Column V: Supporting BRI Participating Countries in Building Green Transportation Infrastructure

1. China-Laos Railway Project

During the construction process of the China-Laos Railway, ecological and environmental protection have been top on the agenda, and environmental protection measures have been stringently implemented so as to build a green and ecological-friendly railway. The project actively carried out environmental impact assessment, and commissioned a local enterprise to compile an EIA report based on the actual situation. To promote the construction of the green corridor, pilots and trial projects have been developed in advance to plant trees, shrubs and grasses in 7 places along the route to compare and select the vegetation suitable for the local area. The total green area of the entire route is about 3.07 million square meters with approximately 8.63 million shrubs and 55,000 trees. To conserve water resources, a total of 157 sedimentation tanks were set up at the tunnel portals along the route. Drilling mud produced in building bridges has been kept in mud tanks and sedimentation tanks for recycling.

2. China-Pakistan Karakoram Highway Realignment Project from Thakot to Raikot

To meet the overall goal of "safety, smooth traffic, environmental protection and conservation", environmental protection, resource conservation, whole-life costs and other requirements have been taken into full consideration when various types of researches were conducted over technical standards, route selection, construction plan comparison and facility construction along the route. Under the premise of ensuring relevant traffic capacity and safety, the project has fully adapted to local terrain and geological conditions so that the highway can harmonize with the natural environment as much as possible. For eco-friendly routing, bridges and tunnels were built to avoid eco-sensitive areas. The project has attached importance to environmental impact assessment work, and commissioned a local enterprise NESPAK to compile the EIA report based on the actual situation. The project strictly follows the local environmental laws and regulations as well as the requirements for environment impact assessment. The opinions from the Pakistan government, local communities and environmentalists have all been incorporated in the feasibility study report and construction plan of the project, winning recognition from the local people for the greening performance of Chinese enterprises.

3. Lome International Airport Expansion and Modernization Project in Togo

The Lome International Airport Expansion and Modernization project in Togo is known as a "model project" in West Africa. It was awarded as one of the China International Sustainable Infrastructure Projects by the China International Contractors Association in 2019. The sustainable development of the airport has been highlighted, and relevant ecological and environmental protection requirements have been stringently observed to create a green and sustainable airport. The project emphasizes reducing greenhouse gas emissions in all stages, including design, construction and operation, and protecting the local ecological environment. For example, an independent IBUS control system was used for the lighting of the terminal's public areas to reduce the heat generated by indoor lights. Under the

the principle of green procurement, strictly select subcontractors and material and equipment suppliers who have passed ISO14000 certification for cooperation the catalog of environmental-friendly building materials and eliminated and restricted building materials have been set up, and green procurement accounts for 95% of the total procurement amount.

2) Promoting green and low-carbon travel

Over recent years, China's new energy vehicle industry has made significant progress in the development of industrial standards, industrial alliances, corporate layout, technology R&D, etc. More and more Chinese new energy vehicle projects have been implemented, from Latin America to Europe and Asia to Africa, to help the BRI participating countries develop intelligent and green public transportation, and promote green and low-carbon production and lifestyle. In 2017, the first electric bus factory built by BYD Company in Europe was put into operation in Hungary, making about 400 zero-emission electric buses every year, thus helping Hungary build a public transportation system with both economic and environmental benefits. By the end of 2022, BYD has delivered 435 electric buses to Chile, accounting for more than 60% of the Chilean electric bus market. It is expected to deliver 800 such buses by the end of 2023, to help Chile achieve its 100% electric bus fleet goal by 2040.

3) Building green logistic and transportation networks

The logistic and transportation networks jointly built by China and BRI participating countries, such as the China-Europe Railway Express and the New Western Land-Sea Corridor Railway Express, have significantly contributed to energy conservation, pollution and carbon emissions reduction. It become a green trade channel connecting the Eurasian continent. By employing the novel organization method for international land-based transport, China-Europe Railway Express has greatly reduced carbon emissions and transportation-induced impact on environment by enhancing international cargo transportation, optimizing the transportation structure, improving efficiency and shortening delivery time. The average carbon emissions of trains account only for 1/7 of road transportation, and it has also driven the rapid development of multi-modal transport. The logistics capacity and transportation processes have been optimized for various transport means, including sea-rail combined transport, road-rail combined transport, air-rail combined transport, etc. The logistic efficiency has thus been improved, making it a model project for tackling global climate change and promoting sustainable development of global transportation.

5. Optimizing Green Financial Cooperation

1) Establishing platforms for green financial development

— **Building a platform for green finance cooperation.** To build a green financial development platform and encourage and guide financial institutions and enterprises to expand green investment is one of the cornerstones of constructing the Green Silk Road. The *Green Investment Principles for the Belt and Road* (GIP) puts forward seven principles including disclosure of environmental information, employment of green financial instruments, and conducting capacity building through multi-party cooperation, etc. A total of 46 financial institutions and enterprises from 17 countries have participated in the GIP, including China's state-owned commercial banks and policy banks, as well as large financial institutions from France, Germany, the United Kingdom, Singapore, Mongolia, Kazakhstan, Thailand and South Africa. The total assets managed or held by its members amount to USD 41 trillion, a regional office for Central Asia and another regional office for Africa have been set up. Based on GIP, an environmental and climate risk assessment toolkit, and a green project database have been set up, and a series of green financial capacity-building activities have been unfolded.

2) Improving green financial services

— Promoting guidance on green investment with policies. China has constantly improved the sustainable development level of its outbound investment and cooperation projects by optimizing green financial policies. Since 2013, a series of policy documents have been issued to guide and regulate the green development of outbound investment projects. *Green Development Guidelines for Foreign Investment and Cooperation* and *Guidelines for Ecological and Environmental Protection in Overseas Investment and Cooperation Construction Projects* have been published. *Guidelines for Country (Region)-Specific Overseas Investment and Cooperation* have been constantly updated. These documents have strengthened support for Chinese enterprises to cooperate with BRI participating countries in green economy, and regulated their environmental performance during overseas investment and cooperation, and to promote the green development of overseas investment and cooperation projects. Besides, China requires enterprises to improve the disclosure of their environmental information in overseas investment, encourages the utilization of green financing instruments like green bonds, and promotes risk management such as compulsory liability insurance for environmental pollution.

— Participating in green investment and financing projects of BRI participating countries. Through policy guidance and financial product innovation, China and BRI participating countries have mobilized more financial resources to green and low-carbon industries, and developed a number of model projects to promote green and low-carbon transformation and upgrading of industries in BRI countries. For instance, the Industrial and Commercial Bank of China provided 13 million dollars of financing support for the Zhanatas Wind Farm in Kazakhstan. This is Kazakhstan's first renewable energy project financed by a commercial bank, and also the country's first attempt to finance a wind power project under project-financing structure. As a global coordinating bank, the Bank of China successfully organized the establishment of an international syndicated facility of several multinational banks and multilateral institutions, including the European Bank for Reconstruction and Development, the Asian Infrastructure Investment Bank, and the OPEC Fund for International Development, which providing crucial financial guarantee for the Syrdarya Gas Power Plant project in Uzbekistan registering over USD 1 billion of investment.

3) Deepening international cooperation on green finance

— Strengthening cooperation with multilateral financial organizations. China has established close ties with multilateral platforms including the Asian Infrastructure Investment Bank (AIIB), World Bank, Asian Development Bank, New Development Bank, Multilateral Cooperation Center for Development Finance, etc., fostering the joint effort to provide financial support to the infrastructure connectivity and green development of BRI participating countries. For instance, China has supported the AIIB to formulate the *Environmental and Social Framework* and *Energy Sector Strategy* in alignment with international standards to guide investors to promote green concepts and technology in infrastructure projects, such as energy, transportation, urban development, etc. By June 2023, the total amount of projects approved by the AIIB reached USD 41.2 billion, with the share of responding to climate change accounting for over 50% of the total. China is committed to further strengthening cooperation with the World Bank and the Asian Development Bank to mobilize more funds to support BRI participating countries in addressing climate change and green and low-carbon development.

— **Fostering green finance standard international cooperation.** In 2018, China and Europe jointly set up the International Platform on Sustainable Finance (IPSF), aiming to deepen international cooperation and mobilize private sector funds to sustainable investment. In June 2022, IPSF officially published the *Common Ground Taxonomy*, which includes 72 economic activities recognized by China and Europe as making major contributions to mitigating climate change. It promotes the mutual recognition and interoperability of green assets between China and Europe, facilitating green investment and reducing the cost of green certification in cross-border transactions.

— **Enhancing green financial capabilities.** Since 2018, Chinese research institutions initiated the Global Green Finance Leadership Program, which has supported the capability building for about 4,500 representatives from over 70 developing countries through on-site and online learning and exchanges on green finance policies and practices. It has effectively helped policy-makers and financial institutions in developing countries to improve green financial capabilities. Chinese financial institutions have also actively supported BRI participating countries to improve the green finance system. For example, they facilitated the communication and discussion between Mongolian financial institutions and Chinese financial authorities, and shared documents such as the *Green Bond Endorsed Projects Catalogue* to support the Mongolian Financial Stability Council to officially approve the National Green Taxonomy, which effectively promotes the development of green investment and financing businesses of its financial institutions.

6. Driving Technological Innovation for Green Development

1) Promoting R&D and sharing of green technology

— Establishing a cooperation mechanism for green technology. China has given full play to the mechanisms such as the Action Plan for Belt and Road Science, Technology and Innovation Cooperation, established joint laboratories on green technology, launched the Belt and Road Special Cooperation Program for Sustainable Development Technology, the Science Action Plan for Sustainable Development, and conducted joint research in this field with other BRI participating countries so as to enhance the scientific and technological support to the Green Silk Road to a larger extent. For instance, China promotes demonstration projects in green technology cooperation. China helped Pakistan with technology research and application in areas such as disaster prevention and reduction, water resources protection and sustainable development. It has also provided support in technological researches for carbon emission reduction in the Philippines, Malaysia, South Africa, etc. China is developing a low altitude shrub grassland project with Ethiopia through researches of key technologies. China has also sorted out a list of green technologies covering the prevention and control of air pollution, water pollution and soil pollution, solid waste treatment, environmental monitoring, green transport and others, to provide technology and experience for BRI participating countries.

— Exploring cooperation methods in green technology. Based on the needs and characteristics of different nations and regions, China has established a sustainable development model focusing on technology transfer and transformation, setting up joint laboratories, and training technical personnel. The China-Myanmar-Mongolia-Congo green mineral joint laboratory, China-Thailand-Malaysia renewable energy joint laboratory, and China-Egypt-Sudan green technology joint laboratory have been set up as transnational technology transfer centers to promote the transfer and transformation of advanced green technologies and to inject innovation into the sustainable development of BRI participating countries. China has established the BRI green technology pool and selected about 40 typical cases with reference and promotion values for wide sharing. These cases, showcasing the current conditions and trends of green development in BRI participating countries, are of great value in promoting the sharing of outcomes made in the green industry.

2)Promoting the exchanges and transfer of green technology

China continuously promotes the building of bases for environmental technology exchanges and transfer, and conducts innovation in cooperation models to promote the exchange and transfer of environmental technologies, and enable Chinese technologies and services to find their way in BRI participating countries. China has established the Belt and Road Environmental Technology Exchange and Transfer Center (Shenzhen) which covers every link on the chain, including incubation, investment, cooperation and transformation of technology, with the business covering more than 60 BRI participating countries. China has set up a flagship green technology platform to connect supply and demand, and 15 green industry joint innovation centers which covering energy conservation and emission reduction, hydrogen energy, safety and energy-saving lighting. China has also promoted the establishment of green technology exchange bases with ASEAN and other nations, so as to leverage the clustering effect of environmental protection industries and promote the implementation of green technology in BRI participating countries.

3) Promoting the alignment of green rules and standards

— **Promoting the formulation of international green standards.** China has supported the International Organization for Standardization (ISO) and the International Electrotechnical Commission (IEC) to establish 9 international standard organizations technical institutions that related to green and low-carbon development. It covers the areas of small hydropower, distributed energy systems and sustainable electrified transport. It also took the lead in formulating and issuing 59 ISO and IEC international standards, including the *General Guidelines for the Selection of Energy Saving Evaluators*. Such efforts have laid a foundation for the coordination of green standards among BRI participating countries.

— **Sharing the green industry standards of China.** China has released 33 national standards for carbon peak and neutrality in foreign languages, as well as multiple green industrial standards in foreign languages. Among those documents, the national standard of *Minimum Allowable Values of the Energy Efficiency and Energy Efficiency Grades for Room Air Conditioners* has been widely applied internationally and has been awarded as one of the global best practices by the United Nations in early 2022. Releasing the foreign-language version of green industry standards can promote the compatibility and alignment of national green standards with BRI participating countries and strengthen exchanges and cooperation in comparing products' energy efficiency standards and coordinating energy efficiency labels, etc.

7. Initiating Active Cooperation on Addressing Climate Change

In 2015, Chinese President Xi Jinping announced at the UN Climate Change Conference in Paris that China would launch cooperation projects to set up 10 low-carbon industiral parks and start 100 mitigation and adaptation programs in other developing countries and provide them witn 1,000 training opportunities on climate change, which is known as the "10-100-1000" Initiative. In 2019, President Xi Jinping announced at the Second Belt and Road Forum for International Cooperation that China would implement the Belt and Road South-South Cooperation Initiative on Climate Change. As of November 2023, China has signed 48 climate change cooperation documents with 39 BRI participating countries and has carried out over 70 projects on climate change mitigation and adaptation with more than 30 BRI countries, including Ethiopia, Pakistan, Samoa, Chile, Cuba and Egypt. Several low-carbon industrial parks have been established to support relevant countries in improving their capabilities on addressing climate change.

1) Jointly building low-carbon industrial parks

China has jointly built low-carbon demonstration zones with Laos, Cambodia and other countries, and by providing materials, such as solar streetlights, new energy vehicles and portable environment monitoring equipment to jointly developed plans for the low-carbon demonstration zones, which showed China's great support to promote the green, low-carbon and sustainable development of BRI participating countries and improved their capabilities to address climate change through the "material + brain power" approach. The low-carbon demonstration zone has strengthened the consensus between China and BRI participating countries on addressing climate change and green and low-carbon transition, and also serves as a valuable experience for these countries to explore low-carbon development paths of their own.

Column VI: Supporting BRI Participating Countries in Building Low-Carbon Demonstration Zones

1. Vientiane Saysettha Low-Carbon Demonstration Zone in Laos

As the first ever low-carbon industrial park jointly built by China and the BRI participating country, the Vientiane Saysettha Low-Carbon Demonstration Zone project is highly consistent with the renewable and green energy development plan proposed by the Lao government. It further drives Laos to achieve its goal of net-zero greenhouse gas emission in 2050, and provides a reference template for overseas industrial parks and comprehensive development zones to explore green and low-carbon transformation and sustainable development. The first batch of assistance supplies for the Laos low-carbon demonstration zone includes 2,000 solar streetlights, 28 new energy vehicles and 5 sets of environment monitoring equipment. The demonstration zone was officially unveiled in April 2022 and is expected to reduce local carbon emissions by approximately 1,243 tons every year, equivalent to planting more than 100,000 trees. The concept of green and sustainable development from China promoted the development of Vientiane, making it a model city in Laos and even among ASEAN countries in terms of low-carbon and green development. Currently, China and Laos have signed the cooperation document on the second batch of assistance supplies for the low-carbon demonstration zone. It will further boost the low-carbon development of the zone.

2. China-Cambodia Low-carbon Demonstration Zone in Sihanoukville

China and Cambodia have jointly built a low-carbon demonstration zone in Sihanoukville Province. The first batch of assistance supplies includes 2,800 solar streetlights, 200 photovoltaic power generation systems for campus, 200 electric motorcycles, 10 automatic air quality monitoring stations, and equipment for climate change response, including 10 sets of portable devices for monitoring industrial flue gas, noise, vibration, etc. "Green" lights have illuminated the Sihanoukville Port, helping Cambodia take a crucial step in addressing climate change. China and Cambodia have signed the Memorandum of Understanding On Cooperation for the Second Batch of Assistance Supplies for the Low-Carbon Demonstration Zone. China will provide Cambodia with 1500 solar streetlights, 2000 photovoltaic power generation systems for households, 2 air quality monitoring vehicles and other equipment for addressing climate change to further support the construction of the low-carbon demonstration zone in Sihanoukville Province.

2) Launching assistance projects in addressing climate change

— **Providing supplies and equipment assistance.** China fully accommodates the actual situation and key concerns of the BRI participating countries on climate change, adhere to tailored assistance programs on local conditions, and provides recipient countries with supplies within its capabilities in the areas of clean energy, low-carbon transportation, disaster early warning and environmental monitoring. For instance, the household photovoltaic power generation system provided by China to Pakistan has greatly improved residents' access to electricity. The Prime Minister of Pakistan remarked that "China's assistance has brought real benefits to the local people". The photovoltaic assistance project for Cuba not only helps them meet the basic needs of electricity and lighting in some remote areas, but also provides new momentum for Cuba to achieve its greenhouse gas emission reduction goal.

— **Supporting engineering project construction.** Besides sending supplies, China also gave strong support in constructing several climate change-related engineering projects, such as the National Emergency Operations Center and its supporting Virtual Emergency Experience Center of Peru, and the Uzbekistan Weather Automatic Station, which help mitigate the influence of climate change locally. Among those projects, the micro-satellite project supported by China in Ethiopia is Ethiopian first artificial satellite, which collects data and give out analysis, helping the country quickly respond to related disasters and further mitigate the damage of climate change on agriculture, forestry, water resources and other related industries. Besides, the two dams built by Chinese enterprises in Tataouine Governorate and Tunisia, have effectively reduced the adverse effects of extreme weather such as heavy rainfall. China also helped Benin repair the dam in Malanville, which improving its ability to control floods. In October 2023, the project approval for the Uzbekistan wind power plant project was completed and put into operation.

3) Initiating training programs in capacity building on responding to climate change

For capacity building, China now has hosted more than 50 training sessions under the South-South Cooperation on Climate Change for 2,300 officials and technicians from over 120 BRI participating countries in the field of climate change, and the sessions are beneficial to them in improving their ability to address climate change. The training programs are mainly aimed at government officials, experts, scholars and technicians working on climate change from the participating countries with a wide range of topics including green, low-carbon and sustainable development, greenhouse gas emission reduction and energy transformation, coping with climate change with aerospace technology, BRI climate finance, low-carbon technology and its related industrial development, ecological adaptation in arid areas, etc. Regional training courses have also been provided for Pacific Island countries and African countries to make the training more targeted through platforms and mechanisms, such as China-Pacific Island Countries Climate Action Cooperation Center. Such efforts have been highly hailed by developing countries. Many countries, such as the Dominican Republic and Nigeria, highly recognize China's contributions in training a large number of local climate professionals, a testament to China's role as a responsible major country with a global vision.

II. Development Experience

Ten years of hard work has yielded fruitful outcomes. China and the BRI participating countries have jointly strengthened the foundation of cooperation, enhanced the impetus for cooperation, expanded the scope of cooperation, and improved the quality of cooperation, pushing green development cooperation to a higher, deeper level in broader areas. Together, the BRI countries and regions have contributed their insights and solutions to building a cleaner and more beautiful world. With 10 years of efforts, they have driven home the concept on green development and consolidated a common understanding of green cooperation. Goals and requirements for green development have been integrated into every aspect and phase of the BRI. Concerted effort has been made to cut carbon emissions, reduce pollution and expand green development while pursuing economic growth. As a result, China and BRI participating countries have achieved coordinated development of economic, social and environmental, and notable progress in building the Green Silk Road.

1. Commitment to the Principle of Achieving Shared Growth through Discussion and Collaboration

The path to green development should be explored based on discussion and collaboration. The humanity lives in a community with a shared future in which benefit to one means benefit to all, whereas harm to one means harm to all. Faced with global ecological challenges like climate change and biodiversity loss, the BRI participating countries adhere to solidarity, cooperation, and common development to learn and benefit each other in pursuing green, low-carbon and sustainable development. The building of the Green Silk Road focuses on vital areas of common concern among participating countries. It emphasizes joint consultations on development topics, collaborative paths and practical solutions. Continuous efforts are made to synergize green development strategies and policies, set goals, directions and tasks, and improve platforms for policy communication, information sharing and technological cooperation. Openness, inclusiveness and sustainability become the keynotes of the BRI with those great efforts.

Green development projects shall be undertaken jointly. China and BRI participating countries explore the potential for cooperation in green development, actively adapted to local conditions, made careful planning, and effectively allocated resources. Efforts are concentrated on bridging the "last mile" for project implementation, exploring opportunities in emerging markets like renewable energy and green industries. Continuous efforts have been made to apply green technologies and implement green projects in BRI participating countries with international advanced production capacity and technologies, providing these countries with practical and feasible solutions for green development. Meanwhile, China innovates international cooperation models, coordinates and integrates differentiated resources, promotes the effective integration of green development needs of BRI participating countries with international advanced production capacity and technologies, and achieves a 1+1+1>3 effect in third-party markets cooperation.

The achievements of green development shall be shared by all. China, as a responsible major country with a global vision, always keeps pace with the times and becomes a leader in global environmental governance. China also deepens its cooperation with BRI participating countries in ecological and environmental protection and sustainable development. The establishment of the China South-South Climate Cooperation Fund and the announcement of the Kunming Biodiversity Fund is to ensure the fruitful result of green development can be enjoyed by all. Grounded in the needs of BRI participating countries for green and low-carbon development, China shares its green philosophies and best practices with participating countries, providing them with holistic and systematic solutions to resolve key environmental problems and improve green benefits.

2. Commitment to High-standard Cooperation to Improve People's Livelihood and Promote Sustainable Development

High standards are constantly and inherently pursued for all matters. High-quality development and high-standard protection are inter-complementary and mutually-reinforced each other. High-standard protection provides vital support to high-quality development. China highly values the ecological and environmental protection of the BRI, consistently raises the environmental standards for ecological and environmental protection in overseas investment and joint construction projects, and encourages the adoption of internationally recognized standards or even more stringent ones. During the implementation process of such projects, China pushes for the use of renewable and clean energy, works towards environment-friendly solutions and greening projects tailored to local conditions, and spurs technological innovation which makes higher standards possible, nurturing new economic growth points so that more opportunities are offered to BRI participating countries in fresh domains and fields.

Improving people's well-being is the fundamental principle. A sound ecological environment is a public good that benefits everyone equally. The Green Silk Road can meet the urgent need to bring tangible ecological and environmental benefits to the people of BRI participating countries and is a crucial means to boost their sense of happiness and security. China attaches great importance to carrying out green livelihood projects and works to enhance our capability of ensuring that green development delivers real benefits to the people. Such benefits include more employment opportunities, a better living environment and increased climate and disaster resilience. With those achievements, the benefits of the BRI now reach all its participants and relevant parties, fostering deeper connections among the peoples of China and its participating countries.

Sustainable development remains the priority goal among all endeavors. Amid the global economic slowdown and the increasing challenges of climate change, China constantly reshapes its developmental advantages and momentum through high-standard environmental protection, offering hope to participating countries grappling with the dual tasks of economic growth and environmental protection. China emphasizes on the sustainability of the Green Silk Road, focuses on areas like green infrastructure, green energy, green transportation, green industries and green trade. It taps into the potential of green finance as a safeguard, striving alongside participating countries to forge a more robust, dynamic and inclusive development path.

3. Commitment to the Principle of Harmonious Coexistence between Human and Nature

Giving priority to ecology. Green and low-carbon growth is the generl direction and trend of current global development and is universally endorsed by all countries. Guided by the principle of green, low-carbon and sustainable development, China and BRI participating countries strive to promote comprehensive green transition of economic and social development. The Green Silk Road and the green, low-carbon goals of respective countries complement and reinforce each other. The green BRI keeps creating numerous highlights of cooperation in such areas as climate change, environmental protection and energy transition. China has given BRI participating countries invaluable green assets by offering them know-how and technological and intellectual support. Those actions demonstrated to the world that China is committed to advancing the building of the Green Silk Road with concreted efforts.

Sticking to actual national conditions. The BRI participating countries are different in development stages and resource endowments. While following general principles, each country must find its own path towards low-carbon green development based on their specific national conditions. A significant majority of those countries are developing ones. Thus, green development in those countries should be pursued in a planned and phased manner with priorities given to certain areas and items. Rich biodiversity is a common feature of BRI participating countries, some of which are typical biodiversity hot spots in the world. Grounded in reality and guided by scientific planning, China and BRI participating countries have initiated various collaborative projects under the Green Silk Road.

Such cooperation actively boosts the capability of ecological conservation and biodiversity preservation.

Adhering to the principle of adopting a holistic approach. The ecosystem is a complex, multidimensional natural environment and a comprehensive social system spanning multiple domains and sectors. Amidst the persisting challenges in global environmental governance, China and BRI participating countries work to better understand and apply the correlation between environmental protection and economic growth when constructing the Green Silk Road, giving priority to addressing prominent environmental issues of BRI participating countries. Also, their needs for industrial restructuring, pollution control, ecological preservation and climate change response are taken into full consideration. Developing countries are supported in their efforts to cut carbon emissions as they pursue development and achieve higher-quality growth by reducing carbon emissions.

Success can only be achieved if one treads on the right path. Building the Green Silk Road is a long-term commitment and a comprehensive project. What was learned from the experiences in past decade is that countries must make comprehensive and persistent efforts to maintain unwavering determination for green development, uphold the principle of shared growth through discussion and collaboration, aim for high-standard cooperation to improve people's livelihood, promote sustainable development, and coordirate resources in various fields. With systematic and persistent efforts, the world will become cleaner and more beautiful.

III. Development Outlook

Green is a prerequisite of sustainable development and an important embodiment of the better life that people desire. Green development is high-quality, sustainable development that follows the laws of nature to promote harmonious coexistence between man and nature. It has become a shared consensus and goal among the BRI participating countries. The Green Silk Road perfectly resonates with the aspirations of the people from those countries for a better life, highly consistent with the UN's Sustainable Development Goals. In this new era, it is sure to thrive with renewed vitality. As long as we press ahead with perseverance, a bright future will beckon. China will work together with participating countries to build a closer green development partnership, peomote policies, rules, and standards linkage and interaction, promote comprehensive and practical cooperation, bolster capacities for green development, and drive the Green Silk Road towards new milestones. This shared journey endeavors to make a new contribution towards establishing a community of humanity and nature.

1. Jointly Boosting Green Development Cooperation

1) Comprehensively solidifying the foundation for green development cooperation

Peace, development, cooperation and win-win are the irresistible historical trend. Over the past decade, the concept and policies of BRI green development have been consistently strengthened. China and the BRI participating countries laid a robust and solid foundation for green development by consistently improving exchange platforms and cooperation mechanisms, and strengthening partnerships and pragmatic collaborations. Moving forward, China stands ready to persistently advance comprehensive cooperation with BRI participating countries, adhering to the principle of shared growth through discussion and collaboration, and aiming for high-standard cooperation to improve people's lives and promote sustainable development. Focusing on ecological protection and green development, China and BRI participating countries will further facilitate successful cooperation projects, and ensure the Green Silk Road delivers more tangible results.

2) Actively expanding the space for green development cooperation

There is tremendous potential and vast scope for cooperation between China and the BRI participating countries in green development. China is ready to collaborate with BRI participating countries to explore avenues around green infrastructure, green energy, green transportation and green finance, leverage platforms such as the BRIGC and the Belt and Road Energy Partnership. Together, we aim to pave the way for green development and build a cleaner and more beautiful world. Additionally, China and BRI participating countries will make great efforts to nurture new growth points in green development cooperation, constantly expanding collaboration in green technology, green industry and green trade. China and BRI participating countries will further enhance the alignment of environment policies, plans and standards, and provide more replicable "small yet beautiful" green solutions. China will give full play to platforms like the BRI Environmental Big Data Platform, the Belt and Road Environmental Technology Exchange and Transfer Center (Shenzhen), promote the transformation of advanced and applicable green technologies, and push forward green technological cooperation networks, green technology innovation and entrepreneurship bases, and international environmental protection industrial parks.

3) Consistently strengthening the capacity building for green development

Capacity building is both the support guarantee for green development and the essential foundation for the BRI participating countries to boost green development. Guided by the philosophy of "teaching a man to fish", China closely collaborating with BRI participating countries, will further enhance capacity building to boost BRI participating countries' internal power for green development based on the actual needs and situations of low-carbon and green growth. China will prioritize areas, such as environmental management, biodiversity and ecological protection, climate change response, green finance, green transport and sustainable infrastructure construction, and carry out capacity building programs, including the Green Silk Road Envoys Program, specialized training sessions on BRI green development, and Capacity-building Alliance of Sustainable Investment (CASI) in a more targeted manner to promote more communications and exchanges among professionals in environmental management and technicians. In addition, China will invest more efforts in talent training in BRI green development, and train a group of versatile ecological protection experts with high caliber and international visions for BRI participating countries, thus building a strong and intellectual support system for green development.

2. Jointly Strengthening Green Development Partnership

1) Deepening multilateral cooperation

China and BRI participating countries will enhance the alignment between the Green Silk Road and South-South cooperation mechanism as well as green development strategies and regional cooperation initiatives of the BRI participating countries in the hope of building broader consensus, sharing concepts and practices of green development, and driving the global environmental governance system towards a more equitable and inclusive direction. Leveraging multilateral agendas, such as the United Nations Framework Convention on Climate Change and the Convention on Biological Diversity, China and BRI participating countries expect to create the synergy among economic development, ecological and environmental protection and climate change response. China will give full play to intergovernmental cooperation platforms, such as the China-ASEAN Environmental Cooperation Center, Lancang-Mekong Environmental Cooperation Center, China Center for SCO Environmental Cooperation, and China-Africa Environmental Cooperation Center, further improve overseas supporting networks for the BRI ecological and environmental cooperation.

2) Enhancing bilateral cooperation

It's important to coordinate the practical needs of BRI participating countries on ecological protection with resilient economic growth in the post-pandemic era in order to build the Green Silk Road. China stands ready to deepen intergovernmental cooperation with those countries, and better align green development strategies, policies, regulations and standards so as to raise the level of cooperation and explore more efficient cooperation mechanisms. China and BRI participating countries will strengthen the research focusing on key topics of green development, promote bilateral dialogues and foster exchanges and cooperation in environmental management systems and environmental protection industries and technologies. China and participating countries will expand cooperation areas, enrich the contents and elevate the levels to ensure more effective results can be delivered. Building on existing BRI cooperation documents and bilateral strategic agreements that signed with BRI participating countries and international organizations, China will promote more pragmatic cooperation under current bilateral environmental cooperation mechanisms. Bearing in mind the needs of BRI participating countries for ecological and environmental protection cooperation, China will work to build more bilateral green low-carbon model projects and share our best practices to more countries.

3) Fostering people-to-people exchanges

Closer people-to-people ties are the cornerstone of effective international cooperation, and the building of the Green Silk Road cannot succeed without a shared aspiration among the peoples. China aims to enhance exchanges and collaboration in the field of green development with non-governmental organizations of the BRI participating countries, and encourage enterprises to implement more green and low-carbon cooperation projects. The goal is to deepen the green consensus among the governments, businesses and the public of the BRI participating countries. Improving the people-to-people exchange mechanism for the BRI green development will serve a bridge for cultural exchanges and closer bonds to connect China with other country. Such a mechanism is expected to deepen the mutual understanding and recognition among the people of participating countries, cultivate a more open and inclusive atmosphere for low-carbon and green cooperation, and set the foundation for public aspiration for embracing a brighter future for the Green Silk Road.

3. Jointly Building a Community of Life for Man and Nature

1) Enhancing ecological protection and pollution control in a coordinated manner

Ecological protection and pollution control is the essential prerequisite for BRI green development and sustainable progress. China stands ready to strengthen strategic alignment and policy exchanges with the BRI participating countries to improve the system of ecological environmental governance as well as ecology and environment management level. To continue improving the ecological and environmental quality, enhanced collaboration will be pursued in the fields of air, water and soil pollution prevention. China and BRI participating countries will promote pollution source control, reinforce the protection and restoration of ecological systems, and make coordinated efforts to reduce carbon emissions, mitigate pollution, expand green development and foster economic growth. China will continue to guide domestic enterprises to enhance their ecological environment management capabilities, instruct them to strictly abide by the ecological and environmental laws, regulations, rules and standards of the host country, and collectively safeguard regional eco-environmental safety.

2) Advancing biodiversity protection in a coordinated manner

Biodiversity is the basis for human survival and development, and it's the blood vein and foundation for the community of life on Earth. China and BRI participating countries will fulfill the obligations made in the Convention on Biological Diversity, and actively take on international responsibilities in line with specific stages of development. Exchanges and cooperation in biodiversity will be deepened by making use of various multilateral and bilateral platforms to promote the implementation of the Kunming-Montreal Global Biodiversity Framework. More efforts will be made to synergize biodiversity conservation and governance so as to usher in a fairer and more equitable global biodiversity governance process during which various countries' strengths can be fully tapped. China will bring into full play its Kunming Biodiversity Fund, and support BRI participating countries in biodiversity-related strategic planning, capacity building, information exchange, and financial project docking to realize the vision of harmonious coexistence between man and nature.

3) Promoting climate change response in a coordinated manner

Climate change is a shared challenge for all human. China aspires to actively seek the greatest possible common ground with BRI participating countries in addressing climate change. China will provide participating countries with feasible financial, technical and capability support under the Green Silk Road, offer solutions for addressing climate change and implementing the UN's Sustainable Development Goals. Continuous efforts will be made to promote the comprehensive, balanced and sustained implementation of the *UN Framework Convention on Climate Change* and *its Paris Agreement.* China will share it's best practices and measures to address climate change and enhance dialogues, exchanges and pragmatic actions to foster a global climate governance system which is equity, justice and win-win cooperation. China will continue to implement the Belt and Road South-South Cooperation Initiative on Climate Change, advance the building of low-carbon demonstration zones, implement climate change mitigation and adaptation projects and provide assistance in combating climate change. Working closely with BRI participating countries, China is committed to improving the capacity to address climate change and jointly protecting our shared home of Earth.

Appendix

Green Silk Road Development Milestones

Year 2015

March 28th: China's National Development and Reform Commission, Ministry of Foreign Affairs, and Ministry of Commerce jointly released *the Vision and Actions on Jointly Building the Silk Road Economic Belt and 21st-Century Maritime Silk Road*, proposing that "Facilities connectivity is a priority area for implementing the Initiative", "Efforts should be made to promote green and low-carbon infrastructure construction and operation management, taking into full account the impact of climate change on the construction", and "We should promote ecological progress in conducting investment and trade, increase cooperation in conserving the ecology and environment, protecting biodiversity and tackling climate change and join hands to make the BRI an environment-friendly one".

September: China announced that it would set up a China South-South Climate Cooperation Fund to provide 20 billion yuan to support other developing countries tackling climate change. Also, China launched the "10-100-1,000" Initiative in 2016, a funding program that aims at financing 10 low-carbon demonstration zones, 100 climate change mitigation and adaptation projects and 1,000 climate professionals training programs to address climate change in developing countries.

November 6th: during the 2015 International Forum on Energy Transitions, the Suzhou Declaration was released, pointing out important areas and directions for energy transformation.

Year 2016

June 17th: The then State Forestry Administration of China and the Secretariat of the United Nations Convention to Combat Desertification jointly released the *Belt and Road Joint Action Initiative for Combating Desertification.*

June 22nd: In a speech made at the Legislative Chamber of the Supreme Assembly of Uzbekistan, Chinese President Xi Jinping proposed that "All partner countries should work together to build a green Silk Road through deepening cooperation in environmental protection, pursuing green development and intensifying ecology and environmental protection efforts".

July 26th: In his congratulatory letter to the 2016 Media Cooperation Forum on Belt and Road, Chinese President Xi Jinping stated that "China is committed to working with countries along the Belt and Road to build a green, healthy, intelligent and peaceful Silk Road that will benefit all the people and countries along the road".

October 30th: During the 2016 International Forum on Energy Transitions, the Suzhou Consensus was released, stating that promoting energy change is an important way to resolve global energy resource and environmental constraints and achieve sustainable development.

December 11th: The Belt and Road Environmental Technology Exchange and Transfer Center (Shenzhen) was officially inaugurated. The Center was jointly established by the Ministry of Environmental Protection of China (current Ministry of Ecology and Environment of China) and the Shenzhen Municipal People's Government. It serves as a physical platform for China and BRI participating countries to engage in international cooperation in environmental protection industries.

December 11th: The proposal of *Honoring Corporate Environmental Responsibilities for Building a Green Belt and Road* was initiated by multiple Chinese enterprises, pledging to comply with environmental regulations, strengthen environmental management in overseas investment and global production capacity cooperation, and contribute to the building of the Green Silk Road.

December: Ministry of Environmental Protection of China (current Ministry of Ecology and Environment of China) signed the *Memorandum of Understanding on Building a Green Belt and Road (2017-2022)* with the United Nations Environment Programme.

Year 2017

April 26th: Ministry of Environmental Protection of China (current Ministry of Ecology and Environment of China), Ministry of Foreign Affairs of China, National Development and Reform Commission and Ministry of Commerce of China jointly released the *Guidance on Promoting Green Belt and Road.*

May 12th: Ministry of Environmental Protection of China (current Ministry of Ecology and Environment of China) released the *Belt and Road Ecological and Environmental Cooperation Plan*.

May 12th: The National Development and Reform Commission and the National Energy Administration of China jointly released the *Vision and Action for Promoting Energy Cooperation on the Silk Road Economic Belt and the 21st Century*

Maritime Silk Road.

May 14th: In the opening ceremony of the first Belt and Road Forum for International Cooperation, Chinese President Xi Jinping proposed that "We will set up a big data service platform on ecological and environmental protection. We propose the establishment of an international coalition for green development on the Belt and Road, and we will provide support to relevant countries in adapting to climate change".

May 15th: A list of deliverables of the First Belt and Road Forum for International Cooperation was released, including *Guidance on Promoting Green Belt and Road*, the *Belt and Road Ecological and Environmental Cooperation Plan*, the establishment of the BRI Environmental Big Data Platform and the setup of the BRI International Green Development Coalition.

June 14th: The BRI Green Supply Chain Platform was officially launched.

November 28th: The Lancang-Mekong Environmental Cooperation Center was established in Beijing.

Year 2018

March 1st: The National Forestry Administration of China issued the *BRI S&T Innovation Action Plan for Ecological Interconnectivity and Mutual Benefits*, proposing the establishment of the BRI S&T Innovation Center for Ecological Interconnectivity and Mutual Benefits to promote green technology exchange.

October 18th: During the the Belt and Road Energy Ministerial Conference and the International Forum on Energy Transitions, China and 17 countries jointly issued the *Joint Declaration on the Establishment of the Belt and Road Energy Partnership*. At the same time, the meeting released *China's Renewable Energy*

Outlook, *China's Renewable Energy Development Report*, *Advanced Power System Flexibility Report*, *China's Urban Energy Report* and other achievements.

November 30th: The Green Finance Committee of the China Finame Society and the City of London jointly initiated the *Belt and Road Green Investment Principles* (GIP).

Year 2019

April 16th: ICBC issued the first green bond of the "the Belt and Road" normalized cooperation mechanism among banks.

April 26th: At the opening ceremony of the Second Belt and Road Forum for International Cooperation, Chinese President Xi Jinping emphasized that "We need to pursue open, green and clean cooperation. The Belt and Road is not an exclusive club, it aims to promote green development. We may launch green infrastructure projects, make green investment and provide green financing to protect the Earth which we all call home", "China and its partners have set up the BRI Sustainable Cities Alliance and the BRI International Green Development Coalition (BRIGC)", "We have set up the BRI Environmental Big Data Platform. We will continue to implement the Green Silk Road Envoys Program and work with relevant countries to jointly implement the Belt and Road South-South Cooperation Initiative on Climate Change".

April 25th: During the Green Silk Road Forum session of the Second Belt and Road Forum for International Cooperation, China and its international partners jointly launched the BRIGC and officially inaugurated the BRI Environmental Big Data Platform.

April 25th: During the Second Belt and Road Forum for International

Cooperation, China and 29 other countries jointly initiated the establishment of the Belt and Road Energy Partnership in Beijing and released the *Cooperation Principles and Concrete Actions of the Belt and Road Energy Partnership*.

April 25th: During the Second Belt and Road Forum for International Cooperation, China's National Development and Reform Commission, the United Nations Human Settlements Programme, World Health Organization and Energy Foundation jointly launched the BRI Sustainable Cities Alliance.

April 27th: A list of deliverables of the Second Belt and Road Forum for International Cooperation was released, including the BRIGC, the BRI Environmental Big Data Platform, the Green Silk Road Envoys Program, the Belt and Road South-South Cooperation Initiative on Climate Change, the BRI Environmental Technology Exchange and Transfer Center, the BRI Green Lighting Initiative, the BRI Green and Efficient Cooling Initiative, the Belt and Road Energy Partnership and the BRI Sustainable Cities Alliance.

November 19th: Ministry of Ecology and Environment of China and Cambodia's Ministry of Environment signed a *MOU on Cooperation in Building Low-Carbon Demonstration Zones* to support Cambodia in enhancing its capability to address climate change.

December 3rd: The first BRI Green Innovation Conference was held in Shenzhen.

Year 2020

July 16th: Ministry of Ecology and Environment of China and Laos' Ministry of Natural Resources and Environment signed a *MOU on Cooperation in Building Vientiane Saysettha Low-Carbon Demonstration Zone* to support Laos in enhancing its capabilities to address climate change.

November 24th: The China-Africa Environmental Cooperation Center was established in Beijing.

December 1st: The BRI Green Development Institute was launched in Beijing.

December 2nd: The Ethiopian Remote Sensing Satellite-1 (ETRSS-1) Delivery Ceremony under China's South-South Cooperation on Climate Change mechanism was held in Beijing, at which a satellite was provided to Ethiopia. It was China's first project which supporting a foreign country to build a satellite, and first remote sensing satellite in collaboration with an African country.

Year 2021

January 9th: The Ministry of Ecology and Environment of China and the Ministry of Agriculture, Climate Change and Environment of the Seychelles signed a *MOU on the Construction and Implementation of a Low-Carbon Demonstration Zone for South-South Cooperation against Climate Change* to support Seychelles in enhancing its capabilities to address climate change.

June 23rd: 29 countries launched the Initiative for Belt and Road Partnership on Green Development at the Asia-Pacific High-level Conference on Belt and Road Cooperation.

June 28th: During the 2021 International Dialogue on Energy Transformations, 49 enterprises jointly released the *Suzhou Initiative in Accelerating Low-Carbon Energy Transition*.

July 9th: The Ministry of Commerce and Ministry of Ecology and Environment of China jointly issued the *Green Development Guidelines for Foreign Investment and Cooperation*.

September 21st: Chinese President Xi Jinping delivered a speech at the General Debate of the 76th Session of the United Nations General Assembly, emphasizing that "We must revitalize the economy and pursue more robust, greener and more balanced global development", "China will step up support for other developing countries in developing green and low-carbon energy and will not build new coal-fired power projects overseas".

October 18th-19th: During the Second Belt and Road Energy Ministerial Conference, such outcomes as *Qingdao Initiative for Belt and Road Green Energy Cooperation*, the *Belt and Road Energy Partnership Charter*, and *Best Practices for International Energy Cooperation* were released under the Belt and Road Energy Partnership.

October 24th: The State Council of China issued the *Action Plan for Carbon Dioxide Peaking Before 2030*, calling for green development of the BRI and further cooperation with BRI participating countries in green infrastructure, green energy and green finance. It also emphasized the role of the BRIGC as a cooperation platform.

October 28th: The Chinese government officially submitted *China's Achievements, New Goals and New Measures for Nationally Determined Contributions* to the Secretariat of the UN Framework Convention on Climate Change, indicating that "With a view to jointly promoting green and low-carbon development and building a green Silk Road, China will actively push forward the implementation of Belt and Road South-South Cooperation Initiative on Climate Change", "On the basis of multilateral and bilateral cooperation platform including the BRIGC, China will foster thematic partnerships on global climate governance and green transition to channel more forces into international cooperation on climate change".

November 10th: During the second China-CEEC Energy Cooperation Forum for Enterprises, the *Ningbo Declaration for Deepening Clean Energy Cooperation between Enterprises of China and CEEC* was released.

Year 2022

January 5th: The Ministry of Ecology and Environment and Ministry of Commerce of China jointly released the *Guidelines for Ecological and Environmental Protection in Overseas Investment and Cooperation Construction Projects.*

January 20th: The Ministry of Ecology and Environment of China and the China Export & Credit Insurance Corporation signed the *Strategic Cooperation Agreement on Jointly Promoting the Construction of Green Silk Road*, in a bid to support the building of a Green Silk Road.

March 16th: The National Development and Reform Commission, Ministry of Foreign Affairs, Ministry of Ecology and Environment and Ministry of Commerce of China jointly issued the *Opinions on Jointly Promoting Green Development of the Belt and Road.*

July 12th: The BRIGC and the Belt and Road Media Cooperation Union jointly launched the *Vista of the Green Silk Road*, an international communication program.

November 9th: The Ministry of Ecology and Environment of China and the Ministry of Environment of Cambodia signed the *MOU on the Second Batch of Material Assistance to Low-Carbon Demonstration Zone for the South-South Cooperation on Climate Change*, continuing to support Cambodia in enhancing its capabilities to address climate change.

November 24th: The 21st Ministerial Meeting of Central Asia Regional Economic Cooperation Program was held. Under the theme of Revitalizing

Regional Cooperation for Green, Sustainable and Inclusive Recovery, China and 9 Central and West Asian countries approved a string of documents, including the *Post-Pandemic Framework for Green, Sustainable and Inclusive Recovery* and the *Central Asia Regional Economic Cooperation Green Energy Alliance.*

December 15th: In his address at the opening ceremony of the High-Level Session of Part II of the 15th Meeting of the Conference of the Parties to the *Convention on Biological Diversity*, Chinese President Xi Jinping pointed out that "We will do our best to provide support and assistance to fellow developing countries through the BRI International Green Development Coalition and the Kunming Biodiversity Fund so as to elevate global biodiversity governance to a new height".

Year 2023

January 19th: The State Council Information Office of China released a white paper titled *China's Green Development in the New Era,* highlighting that China remains firmly committed to green development of the BRI, ensuring that green remains the defining feature of the BRI.

April 12th: During the Belt and Road Forum on Early Warnings for All, the Ministry of Ecology and Environment of China, the China Meteorological Administration and the World Meteorological Organization signed a tripartite cooperation agreement to support the United Nation's Early Warnings for All Initiative.

May 10th: The first General Assembly of BRIGC was held in Beijing. BRIGC became the first international organization established under the Green Belt and Road Initiative.

May 19th: Chinese President Xi Jinping presided over the first China-Central

Asia Summit and delivered a keynote speech, in which he emphasized that "It is imperative that we continue to set the pace for Belt and Road cooperation and deliver on the Global Development Initiative. We should fully unlock our potentials in traditional areas of cooperation, forge new drivers of growth in poverty reduction as well as green and low-carbon development, and work together to build our community into one featuring win-win cooperation and common progress".

May 24th: During the third Belt and Road Energy Partnership Forum, the initiative on *BRI Energy Transition Supported by Green Finance* was released.

August 3rd: The Ministry of Ecology and Environment (National Nuclear Safety Administration) of China and Pakistan Nuclear Regulatory Authority signed the *MOU on the Establishment of the China-Pakistan Nuclear Safety Cooperation Center*.

September 6th: During the 2023 International Forum on Energy Transitions, China released the *Global Clean Energy Cooperation Partnership*. 43 Chinese and international energy companies and research institutions jointly launched the *Initiatives to Prepare for the International Energy Change Alliance*, establishing a national energy change quantitative evaluation system for the first time and releasing the *Blue Book of Energy Change Index*.

September: The Ministry of Transport of China signed the *MOU on Promoting Sustainable Transport for the SDGs* with the United Nations.

September 11th: The Bank of China issued the first Belt and Road green bond.

October 18th: In his keynote speech at the Opening Ceremony of the Third Belt and Road Forum for International Cooperation, President Xi Jinping stated that China will continue to deepen cooperation in areas, such as green infrastructure, green energy and green transportation, and step up support for the BRI International Green Development Coalition. China will continue to hold the BRI Green Innovation Conference, and establish dialogue and exchange mechanisms for the solar industry and a network of experts on green and low-carbon development. China will implement the Green Investment Principles for the Belt and Road, and provide 100,000 training opportunities for partner countries by 2030.

October 18th: Ministry of Ecology and Environment and the National Development and Reform Commission co-hosted the Third Belt and Road Forum for International Cooperation High-level Forum on Green Development. The *Beijing Initiative for Belt and Road Green Development*, *Green Investment and Finance Partnership (GIFP)*, and the *Action Plan for Green Technology Development in Central Asia* were launched.

October 18th: the Ministry of Commerce hosted the Third Belt and Road Forum for International Cooperation Thematic Forum on Trade Connectivity . China and 34 countries jointly released the *Initiative on International Trade and Economic Cooperation Framework for Digital Economy and Green Development.*